Cambridge Physical Tracts

COSMIC RAYS AND MESOTRONS

CAMBRIDGE
UNIVERSITY PRESS
LONDON: BENTLEY HOUSE
NEW YORK, TORONTO, BOMBAY
CALCUTTA, MADRAS: MACMILLAN
TOKYO: MARUZEN COMPANY LTD

COSMIC RAYS AND MESOTRONS

by

H. J. J. BRADDICK, PH.D.
Birkbeck College, London

CAMBRIDGE
AT THE UNIVERSITY PRESS
1939

CAMBRIDGE UNIVERSITY PRESS
Cambridge, New York, Melbourne, Madrid, Cape Town, Singapore,
São Paulo, Delhi, Dubai, Tokyo, Mexico City

Cambridge University Press
The Edinburgh Building, Cambridge CB2 8RU, UK

Published in the United States of America by Cambridge University Press, New York

www.cambridge.org
Information on this title: www.cambridge.org/9780521178570

First published 1939
First paperback edition 2011

A catalogue record for this publication is available from the British Library

ISBN 978-0-521-17857-0 Paperback

GENERAL PREFACE

It is the aim of these tracts to provide authoritative accounts of subjects of topical physical interest written by those actively engaged in research. Each author is encouraged to adopt an individualistic outlook and to write the tract from his own point of view without necessarily making it "complete" by the inclusion of references to all other workers or to all allied subjects; it is hoped that the tracts may present such surveys of subjects as the authors might give in a short course of specialised lectures.

By this means readers will be provided with accounts of those subjects which are advancing so rapidly that a full-length book would be out of place. From time to time it is hoped to issue new editions of tracts dealing with subjects in which the advance is more rapid.

M. L. O.
J. A. R.

CONTENTS

AUTHOR'S PREFACE

During the last two years, the application of quantum theory to cosmic rays has made it possible to give a connected account of a large part of the phenomena. The theory of radiation from a moving electron has provided a satisfactory account of the behaviour of the less penetrating cosmic rays, and it is now clear that the more penetrating rays cannot be explained in terms of electrons. It has become necessary to introduce a new particle to account for these rays, and its properties are not yet fully explored.

In the present account, the first chapters are devoted to the experimental facts of cosmic rays, and the later to their explanation in terms of the properties predicted by quantum theory for moving charged particles.

The question of the *origin* of the rays has not been considered.

H. J. J. B.

January 1939

Chapter I

INTRODUCTION

When a closed vessel is shielded as far as possible from the radioactive radiation of its surroundings, there is a residual production of about 2 ions per sec. per c.c. of air. The first definite evidence of the cosmic radiation was obtained by balloon flights which showed that this residual ionisation increased with altitude and might therefore be attributed to a penetrating radiation entering the earth's atmosphere from outer space (1). About 1926, Millikan and his collaborators took up the study of the rays, and made measurements of the absorption in air and water. Assuming that the radiation was electromagnetic and of γ-ray type, they interpreted their results on the basis of the then available formulae for X-ray absorption and assigned to the rays quantum energies of the order 10^8 electron-volts. D. Skobelzyn (2) found in a Wilson cloud chamber β-ray tracks whose energies, measured by their curvature in a magnetic field, went up to $1{\cdot}5 \,.\, 10^7$ e.v. Bothe and Kohlhörster (3) interposed a block of gold between two Geiger-Müller β-ray counters and showed by observing coincident discharges of the two counters that there were corpuscular rays whose penetrating power was of the same order as that measured for the cosmic radiation. This work suggested the possibility that the primary radiation was corpuscular, and Bothe and Kohlhörster pointed out that charged particles approaching the earth from outer space should be deflected by the earth's magnetic field. The field would prevent the less energetic particles from reaching equatorial regions, and give rise to a variation of cosmic-ray intensity increasing from the equator toward the poles. A variation of this kind was found by Clay and extensively explored by Compton. This latitude variation shows that a large part of the radiation consists of charged particles; the most recent calculations, based on high-altitude measurements in different latitudes, indicate that the rays reaching the top of the atmosphere are almost entirely corpuscular (4). When the rays enter the atmosphere there is a complicated process of absorption and secondary ray production, and it is to be emphasised that the deflection by the earth's magnetic field takes place outside the atmosphere and gives direct evidence about the primary incident particles.

The energy distribution of the particles which reach the earth's surface may be investigated by the curvature of the tracks in a strong magnetic field. The particles are found to have energies up to the limit observable at about $7 \cdot 10^{10}$ e.v. There is indirect evidence that there exist particles of much greater energy than this. The tracks observed in the expansion chamber indicate particles of both signs, and these tracks in fact provided the first evidence for the positive electron (positron).

Amongst the interactions of cosmic particles with matter, a most characteristic phenomenon is the production of showers. In Skobelzyn's early cloud-chamber experiments there were a few observations of the simultaneous passage of several tracks through the chamber. Further work with the Wilson chamber, and with sets of Geiger counters, showed that showers of particles were found most frequently in the neighbourhood of heavy elements. In chamber photographs of typical showers, a number of tracks came from a limited region of the heavy element, or from a number of such centres. In experiments with ionisation chambers there had been found occasional bursts of ionisation, corresponding to the simultaneous production of 10^5–10^7 ion pairs. These "bursts" or "Stösse" were first thought to be a separate phenomenon, but it now seems probable that most of them are very large showers of the type described above. Recently, the theoretical importance of the showers has become evident, for it appears that the quantum theory, if applied to electrons and γ-rays of very high energy, in a region where its validity was formerly uncertain, predicts the production of showers by a cascade process.

Some of the cosmic particles, primary or secondary, have very great penetrating power, for cosmic rays can be detected, not only at the bottom of the normal atmosphere, but under great depths of water and in deep mines. Absorption measurements show that the radiation at sea level can be split rather definitely into a more absorbable and a less absorbable component. The more absorbable component may be satisfactorily interpreted as consisting of electrons and photons, which lose energy according to quantum theory. The hard component is very much more penetrating. It is likely that the hard component consists of particles with hitherto unknown properties, and there is a considerable body of evidence that they have a mass intermediate between the electron and the proton masses. It appears probable that these particles are secondary to incoming electrons, rather than primary particles from outside.

REFERENCES

(1) Hess, *Phys. Zeit.* **13**, 1084, 1912; *Wien Ber.* **121**, 2001, 1912.
(2) Skobelzyn, *Z. Physik*, **54**, 686, 1929.
(3) Bothe and Kohlhörster, *Z. Physik*, **56**, 751, 1929.
(4) Bowen, Millikan and Neher, *Phys. Rev.* **53**, 855, 1938.

Chapter II

METHODS OF INVESTIGATION

2·1. In modern cosmic-ray investigations the main methods used are:

(1) The ionisation chamber in which the ionisation current is measured by an electroscope or electrometer.

(2) Systems of Geiger-Müller counters.

(3) The cloud chamber of C. T. R. Wilson.

As a special method may be mentioned the photographic emulsion technique of Blau, in which the tracks of an ionising particle are observed by the lines of blackened grains left in a photographic emulsion. This method has been applied to the discovery of rare disintegrations producing particles of mass comparable with α-particles or protons.

2·2. The ionisation chamber.

An ionisation chamber for cosmic-ray measurement is usually filled with gas under pressure; an electric field is applied to drive the ions to a collecting electrode, and the ionisation current is measured by some form of electrometer. Some of the ions are lost by recombination before they are collected. This loss depends on the collecting electric field, the pressure and nature of the gas, and the distribution of the ionisation about the electrode system. The loss by recombination is larger for a heavily ionising particle than for an electron, and the variation of ionisation with pressure has been used to provide evidence that the particles in bursts ionise like electrons.

Recombination is less marked in pure argon than in most other gases, and the specific ionisation along an electron track is large in this gas. It is therefore often used in cosmic-ray measuring chambers. The ionisation chamber is valuable in accurate comparisons of cosmic-ray intensity, since the measuring arrangement has stable characteristics and the calibration may be checked with a radioactive source. It may be used in the study of the geographical and time variations of the intensity.

In the form of a light electroscope carried by balloons, the ionisation chamber has been used for investigating the change of cosmic-ray intensity with altitude. In the interpretation of these results, however, there is a complication due to the fact that rays are incident at all angles and reach the instrument through different thicknesses of air. The theoretical investigations of the passage of rays through absorbing layers lead naturally to an absorption curve for unidirectional radiation. Such a curve is obtained directly by a vertical counter arrangement, but Gross (1) has shown that it may be obtained from ionisation-chamber measurements by the transformation

$$\psi(x) = J(x) - x\frac{dJ(x)}{dx},$$

where x = depth of penetration,

$J(x)$ = observed intensity (all directions),

$\psi(x)$ = computed intensity of a normally incident beam.

In deducing this formula it is assumed that the initial rays enter isotropically over a hemisphere and that the absorption and secondary production of a particular ray is confined to a single line, i.e. that there is no sideways scattering.

2·3. Superimposed on the steady ionisation current in an ionisation chamber there are bursts of ions due to the simultaneous passage of a number of ionising rays. These may be measured by a recording electrometer or by a valve amplifier which responds to the sudden increase in ionisation current. The lower limit to the size of a detectable burst is set by fluctuations in the ordinary ionisation current due to unrelated cosmic rays arriving at random times. If a number of such rays happen to pass in a time comparable with the time of collection of the ions, or the time of response of the recorder, a spurious burst will be recorded. In the case of an ionisation chamber connected to a valve amplifier it is usually the time of collection of ions which must be considered in this connection; when an electrometer is used it is frequently the time of swing of the moving needle. Formulae for calculating the number of spurious bursts due to fluctuations may be obtained (2). For the case where the average number of particles arriving in a time equal to the "time of resolution" is small compared with the least number of simultaneous particles counted as a burst, the number of spurious bursts is given by an approximate formula:

$$P = \frac{1}{T}\sqrt{\frac{n}{2\pi}}\,e^{-k^2/2n},$$

where P = number of "spurious bursts" with more than k particles,

m = average number of random particles per second,

T = time of resolution,

$n = mT$

$k \gg n.$

In the case of a small chamber used by R. T. Young (3), which may be taken as typical, bursts of ten rays could be recorded, and the number of spurious bursts of this size, due to fluctuations, was negligible.

2·4. The Geiger-Müller counter.

The Geiger-Müller counter usually takes the form of a tubular cathode and a fine wire anode sealed up in a tube containing gas at low pressure. If the voltage applied to the counter lies within certain limits, the ionisation produced by the passage of a particle through the counting tube is magnified by collision and other processes within the counter itself. Each particle gives rise to a short quenched discharge, which ordinarily lasts for about 10^{-2} sec. This time may be reduced by the use of special quenching circuits. A counter tube responds to the passage of cosmic-ray particles, but it also responds to radioactive contamination and most tubes probably give also spontaneous discharges. There are a number of circuits which record only the coincident discharges of a number of counters (4)—discharges being coincident if they occur within a certain time called the resolving time of the apparatus. In practice, with simple circuits, this time is of the order 10^{-3} sec., and it can be reduced by the use of special circuits. The resolving time can be made less than the duration of a single counter impulse because the amplifying circuits can be adjusted to respond only to the initial, steeply rising, part of the impulse. Local radioactivity does not give rise to genuine coincidences, since radioactive β particles cannot in general penetrate more than one counter, but there are apparent coincidences due to the random passage of a particle in each of the counters within the resolving time. The number of these casual "coincidences" in unit time is given in the case of two counters by

$$A_{12} = 2N_1N_2\tau,$$

where $N_1 N_2$ = average numbers of random impulses in the counters in unit time,

τ = resolving time.

The corresponding approximate formula for the case of three counters is

$$6N_1N_2N_3\tau^2.$$

The number of casual coincidences in practical cases decreases very rapidly with an increase in the number of counters.

In the simplest use of the coincidence system, two or more tubular counters are arranged with their axes parallel (Fig. 2·1 *a*). The passage of a single particle through all the counters produces a count, and the arrangement may be used as a "telescope" to distinguish rays coming from a particular direction. A set of counters arranged in this way may be used to measure the absorption of the rays, the blocks of absorbing material being interposed between the counters (§ 4·3).

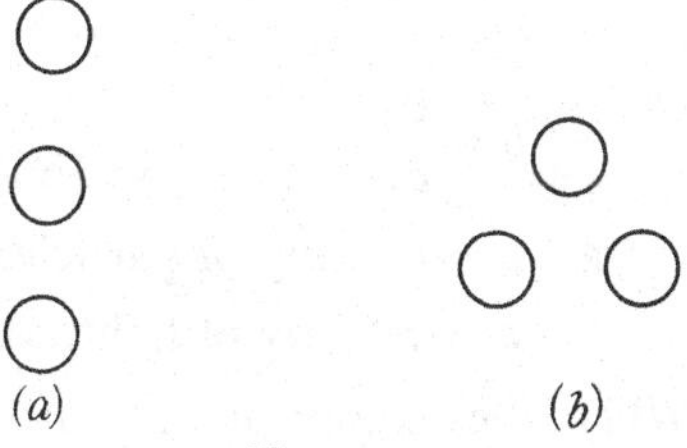

Fig. 2·1.

The method of coincidence counting has been extended to the observation of more complicated events. The arrangement of counters shown in Fig. 2·1 *b* responds only if at least two particles pass simultaneously, and it is typical of a number of arrangements used in the investigation of showers.

2·5. The Wilson cloud chamber.

In the cloud chamber, the tracks of ionisation left by individual particles are made visible by the condensation of liquid drops on the ions as nuclei.

The uses of the cloud chamber in cosmic-ray research are of three main types:

(1) The chamber may be used to show the general nature of the phenomena and for the statistical study of some special phenomenon such as shower production.

(2) The chamber may be used in a magnetic field and the curvature of the tracks measured. This allows a direct deduction of the energy of the particles, for we have

$$H\rho = \frac{m}{e}\frac{\beta c}{\sqrt{1-\beta^2}},$$

$$10^8 V = \frac{mc^2}{e}\left(\frac{1}{\sqrt{1-\beta^2}} - 1\right),$$

where H = magnetic field (gauss),

ρ = radius of curvature,

V = energy in electron-volts,

$\beta = \frac{v}{c}$,

v = velocity of particle,

c = velocity of light.

When the energy of the particle is very large compared with mc^2, we have

$$V = H\rho\frac{c}{10^8} = 300H\rho,$$

and the mass of the particle does not appear in the expression.

For a particle of energy 10^9 e.v. and a field of 10,000 gauss, the radius of curvature is 3 metres, so that refined curvature measurements are necessary to measure the energy of fast particles. In particular the distortion of the tracks, due to irregular motion of the air in the chamber, must be reduced to a minimum. At the present level of technique, the energy of a particle of 10^9 e.v. can be measured to about 5 per cent and the curvature can just be detected for a particle of energy $7 . 10^{10}$ e.v.

(3) The chamber photographs may be used to find the density of ionisation along the track (the "specific ionisation"). It will be shown (§ 6·2) that in a certain energy range the density of ionisation of a particle of given energy depends on the mass of the particle. Measurement of the specific ionisation and the curvature of the track of a particle in this range enables its mass to be determined. For the purpose of determining the ionisation the chamber conditions may be set to give diffuse tracks in which individual droplets, corresponding to single ions, can be counted.

Cosmic-ray tracks were first found in chamber photographs taken for a different investigation, and the expansions were made at random times as far as cosmic rays were concerned. The probability of finding a cosmic-ray track in a random expansion depends on the time during which the gas in the chamber is in a condition to deposit droplets on ions. The considerations affecting this will be discussed below. With a chamber of ordinary design, Anderson obtained one track in about five expansions, but only a small proportion of these were suitable for detailed study. Blackett and Occhialini (5) introduced a technique for making the passage of the ray take its own photograph, the expansion being initiated by the simultaneous discharge of Geiger counters fixed above and below the chamber. This method gives a large yield of tracks which are suitable for accurate curvature measurement, but it must be remembered that the events photographed are selected according to the probability of their tripping the counters. Showers, for example, occur more frequently than in random expansions, and particles which come to the end of their range in the chamber are never photographed at all.

The use of a cloud chamber for cosmic-ray recording is governed by the following considerations. The gas in the chamber contains vapour (usually water or alcohol) which becomes supersaturated when the gas is cooled by a sudden expansion. In this supersaturated state, droplets condense on any ions present. If a particle passes through the chamber before the expansion, the ions begin at once to diffuse, and they spread until fixed by the deposition of droplets. Blackett calculates that after $\frac{1}{70}$ sec. 90 per cent of the ions in a track lie inside a width of 1 mm., and that the width of a track increases with the square root of the time. Old tracks have therefore a considerable breadth and cannot be used for accurate measurement of curvature. It is, however, possible to count the droplets in such a diffuse track, and for this purpose the expansion of a counter-controlled chamber may be artificially delayed. The diffusion may also be increased by the use of hydrogen in the chamber, on account of the greater mobility of ions in that gas.

It is clear from the data above that tracks which are more than a few tenths of a second old when the expansion takes place will be unrecognisable. After the expansion, the gas in the chamber remains supersaturated for some time, and therefore sensitive to ionising particles. This sensitive period is terminated by the gradual warming up of the chamber, and it may be lengthened to the order of 1–2 sec. by using a large and carefully designed chamber. E. J. Williams has obtained about one useful cosmic track per random expansion by using this period.

REFERENCES

(1) Gross, *Z. Physik*, **83**, 214, 1933.
(2) Ehrenberg, *Proc. Roy. Soc.* **155**, 532, 1936.
(3) Young, *Phys. Rev.* **52**, 559, 1937.
(4) For theory and practice of coincidence counting see, e.g., Rossi, *Z. Physik*, **82**, 151, 1933; Johnson and Street, *Journ. Frank. Inst.* **215**, 239, 1933; Street and Woodward, *Phys. Rev.* **46**, 1029, 1934.
(5) Blackett and Occhialini, *Proc. Roy. Soc.* **139**, 699, 1933; Blackett, *Proc. Roy. Soc.* **146**, 281, 1934.

Chapter III

THE VARIATION OF COSMIC-RAY INTENSITY WITH TIME AND LATITUDE

3·1. Time variation of the cosmic-ray intensity.

The intensity of the cosmic radiation at a given station is very constant in time. Many attempts have been made to detect small periodic variations corresponding to the solar or the sidereal day. The experimental problem is complicated by barometric changes which cause variations in the screening effect of the atmosphere. No variations in excess of about 0·2 per cent have been detected, so that the cosmic radiation in the neighbourhood of the earth is apparently highly isotropic. There is, however, no direct information about the initial distribution of the rays, because the particles are charged and their motion may be affected by electric and magnetic fields in space. There is known to be an excess of positive incident particles, at least up to an energy of about $2 \cdot 10^{10}$ e.v., and it has been pointed out that these constitute a space charge which is probably neutralised by slow-moving ions. Any relative motion of the cosmic-ray distribution and the ion distribution would set up a magnetic field which would bend the paths of the particles. Alfvén (1) has discussed this effect, and considers that it prevents the appearance of a sidereal time variation due to particles of extra-galactic origin.

3·2. The latitude effect.

The cosmic-ray intensity in equatorial regions is about 16 per cent smaller than in Northern Europe. The results of a systematic survey show that the cosmic-ray intensity is nearly a function of geomagnetic latitude (2). The geomagnetic latitude is the latitude referred to the axis of the "equivalent dipole" which represents the regular part of the earth's magnetism, and it is the simplest angular coordinate for the earth's magnetic field at a distance from the earth remote from local anomalies.

The equatorial decrease in cosmic-ray intensity is clearly due to the effect of the magnetic field on charged particles approaching the earth. An analysis of the motion of charged particles in the field of a dipole has been made by Störmer in connection with the theory of the aurora, and applied to cosmic rays by Rossi, Lemaître and Vallarta, and others (3). The particles of given energy V e.v. have a "magnetic stiffness" $H\rho$ which is given, for very energetic particles with unit charge, by the expression

$$H\rho = \frac{V}{300} \quad \text{(see § 2·5)}.$$

If M is the magnetic moment of the dipole, we shall write

$$k^2 = \frac{M}{H\rho} = \frac{300M}{V}.$$

When the particles move in the field of the dipole the magnetic acceleration is always normal to the direction of motion, so that the particles move with constant velocity and constant relativistic mass.

By integrating the equations of motion, Störmer obtained the equation

$$\sin\theta = \frac{k^2 \cos\lambda}{r^2} - \frac{2\gamma}{r\cos\lambda},$$

where r, λ are the radial distance and latitude which define the position of the particle with respect to the dipole,

θ is the inclination of the trajectory to the plane of the magnetic meridian,

γ is a constant which depends on the initial conditions and varies from $-\infty$ to $+\infty$ if we take all initial directions as possible.

It is convenient to write

$$\sin\theta = \frac{\cos\lambda}{x^2} - \frac{2\gamma_1}{x\cos\lambda}, \tag{1}$$

where

$$x = \frac{r}{R_0}\sqrt{\frac{V}{V_0}},$$

$$R_0 = \text{radius of the earth},$$

$$V_0 = \frac{300M}{R_0^2} = 6.10^{10} \text{ e.v.},$$

$$\gamma_1 = \frac{\gamma}{R_0} \frac{V}{V_0}.$$

Since $-1 < \sin\theta < 1$, equation (1) involves a division of space into regions accessible to particles of a given velocity and initial conditions, and regions for which $\sin\theta > 1$, showing that they are not accessible for such particles (Fig. 3·1). These regions are divided

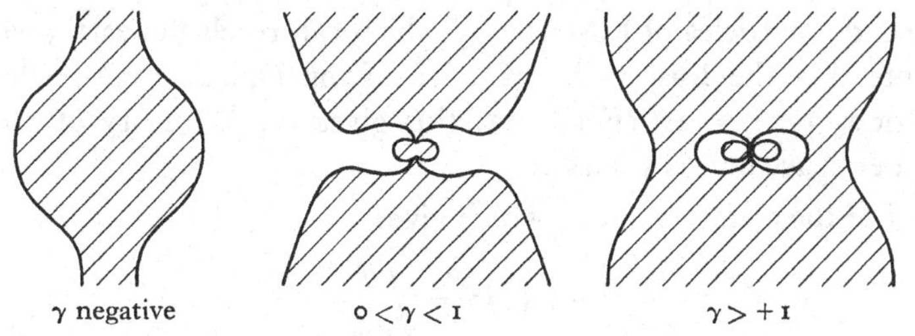

Fig. 3·1. Meridional section of the forbidden regions of Störmer's theory. The magnetic dipole is vertical at the centre of each figure.

by surfaces (lines in the meridional plane represented in Fig. 3·1) for which

$$\frac{\cos\lambda}{x^2} - \frac{2\gamma_1}{x\cos\lambda} = \pm 1. \tag{2}$$

The roots of this equation are

$$x_1 = \frac{\gamma_1}{\cos\lambda} + \sqrt{\frac{\gamma_1^2}{\cos^2\lambda} - \cos\lambda},$$

$$x_2 = \frac{\gamma_1}{\cos\lambda} - \sqrt{\frac{\gamma_1^2}{\cos^2\lambda} - \cos\lambda},$$

$$x_3 = -\frac{\gamma_1}{\cos\lambda} + \sqrt{\frac{\gamma_1^2}{\cos^2\lambda} + \cos\lambda},$$

$$x_4 = -\frac{\gamma_1}{\cos\lambda} - \sqrt{\frac{\gamma_1^2}{\cos^2\lambda} + \cos\lambda}.$$

Of these, x_4 is always negative and may be disregarded.

When γ_1 is negative, x_3 is positive for all values of λ: x_1, x_2 are negative, and there is a single bounding surface.

When γ_1 is positive and < 1, there is a single positive real root for $\cos^3 \lambda > \gamma^2$, and three positive real roots for other values of λ.

When γ is positive and > 1, there are three real roots for all values of λ, and the regions of space for which $|\sin\theta| < 1$ are multiply connected. The inner space is not accessible to electrons coming from infinity.

As the particle approaches the earth, x decreases until it reaches the value $x = \sqrt{(V/V_0)}$ at the earth's surface. In order to find the particles of least energy which can reach the earth, we note that the least value of x satisfying equation (2) is the root x_3 for $\gamma = 1$. (For $\gamma < 1$, this gives the boundary of the inner space and is useless.)

For the equator, $\lambda = 0$, which gives

$$x = -1 + \sqrt{2} = \sqrt{\frac{E}{E_0}},$$

$$E = E_0(\sqrt{2} - 1)^2 = 1 \,.\, 10^{10} \text{ e.v.}$$

Particles with this energy can reach the earth at the equator only when coming in a certain direction. For positive particles, this direction of easiest access is from the west. For a particle coming from the east, the least energy required to reach the earth is obtained from x_2 and is given by

$$x = x_0,$$

$$E = E_0 = 6 \,.\, 10^{10} \text{ e.v.}$$

At any other latitude, the minimum allowed energy is obtained from x_3, and corresponds to access (for a positive particle) from the west.

Lemaître and Vallarta showed that for energies greater than this minimum the access of particles is confined to a cone of rather complicated shape bounded by trajectories of two kinds:

(1) Those asymptotic to periodic orbits—trajectories lying just beyond these are returned to outer space without touching the earth.

(2) Those which are tangential to the earth's surface at points other than that under consideration.

The shape of the cone of allowed radiation has been calculated by setting up the equations of motion on a differential analyser and tracing orbits until the asymptotic ones were found. At any given latitude, the cone for positively charged particles of the minimum allowed energy is low on the western horizon, and as the energy of the particles is raised the cone expands until it fills

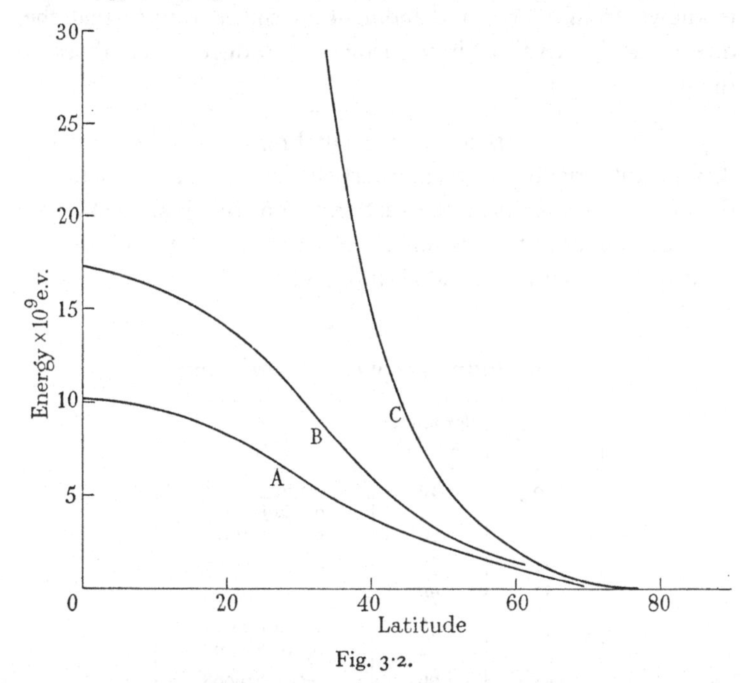

Fig. 3·2.

the whole hemisphere of the sky. Fig. 3·2 shows for each latitude the minimum energy for a particle to reach the earth in the most favourable direction (A), the minimum energy for a particle to reach the earth arriving from the zenith (B) and the minimum energy allowing the rays to arrive in any direction (C).

Assuming that space remote from the earth contains cosmic particles with velocities distributed isotropically, it has been shown that the application of Liouville's theorem leads to the conclusion that the particles in an "allowed" cone are also distributed isotropically, and that their density (the number per sec. per unit area per unit solid angle) is the same as that of the same particles in outer space.

Therefore if the radiation is initially isotropic, the calculation of the allowed cones is sufficient to determine the angular distribution of the radiation at the top of the atmosphere. The rays observed in the lower part of the atmosphere are mainly secondary, but it is known from energy and momentum considerations that the directions of arrival of these particles reproduce closely those of the primary rays.

3·3. The most striking experimental result of a survey of the direction of arrival of the cosmic particles is an asymmetry between the number of rays from the east and from the west. Table 3·1 gives the values of the asymmetry observed by T. H. Johnson at some different latitudes and altitudes (4).

TABLE 3·1

East-west asymmetry at 45° zenith angle

Latitude	Barometer cm. Hg	$A=\frac{I_W-I_E}{\frac{1}{2}(I_W+I_E)}$
0	46	0·139 ± 0·008
	52	0·140 ± 0·016
	76	0·145 ± 0·016
20	76	0·066 ± 0·007
28	76	0·023 ± 0·023
49	46	0·016 ± 0·007
	52	0·019 ± 0·008
51	76	0·005 ± 0·008

The results show a considerable excess of particles coming from the west in equatorial latitudes. Since this is the direction of easiest access for positively charged particles, the east-west asymmetry indicates an excess of primary positive particles in the energy range corresponding to an incomplete cone of access.

The complete analysis of these results is complicated by the atmospheric absorption, which is itself a function of zenith angle.

It is probable that there is a considerable excess of positive over negative incoming particles, and according to Johnson it is possible that all the latitude-sensitive rays, i.e. all the primary particles in the energy range $3\text{–}17 \cdot 10^9$ e.v., are positively charged.

3·4. The account given above of the effect of the earth's field is appreciably simplified (5). Exact calculation of the allowed cones shows that an asymmetry is to be expected between the rays coming from the north and south, and this has been detected experimentally. Considerable deviations of the cosmic-ray intensity from strict dependence on geomagnetic latitude have been ascribed to the fact that the earth's magnetic field is not that of a single symmetrically placed dipole. This gives rise to a "longitude effect", e.g. at different points on the magnetic equator the cosmic-ray intensity is a periodic function of longitude and varies over a range of about 8 per cent. This effect can be explained by the theory of allowed cones with the assumption of a dipole at the magnetic centre of the earth, some distance removed from the centre of the geoid.

3·5. The latitude effect in high latitudes.

The intensity of the cosmic rays at sea level shows an increase from the equator to about latitude 50°, and remains constant at higher latitudes (Fig. 3·3). At latitude 50° particles with energy $3 \cdot 10^9$ e.v. can penetrate the earth's magnetic field and arrive vertically at the top of the atmosphere. The absence of a latitude variation at higher latitudes indicates either that there are no incident particles with energies less than this, or that less energetic particles cannot make their effects felt through the atmosphere. This problem may be resolved by measurements at high altitude. Cosyns (6) drifted in a balloon between latitudes 47° N. and 51° N. at levels where the pressure was 70–180 mm. Hg; he found that the increase of intensity with latitude ceased at about 50° as at sea level. This result has been confirmed by recent high-altitude measurements near the magnetic pole (7), and it is consistent with the work of Bowen, Millikan and Neher described below (§ 4·2).

It therefore seems that the incoming energy spectrum of cosmic rays does not extend below about $3 \cdot 10^9$ e.v., and it has been suggested (8) that it is cut off by the magnetic field of the sun. This field may act on cosmic-ray particles so as to prevent rays of less than a certain energy from reaching the neighbourhood of the earth's orbit, and so as to restrict to certain directions the access of particles over a further energy range. The theory is exactly similar to that given above for the earth. It is not quite clear if

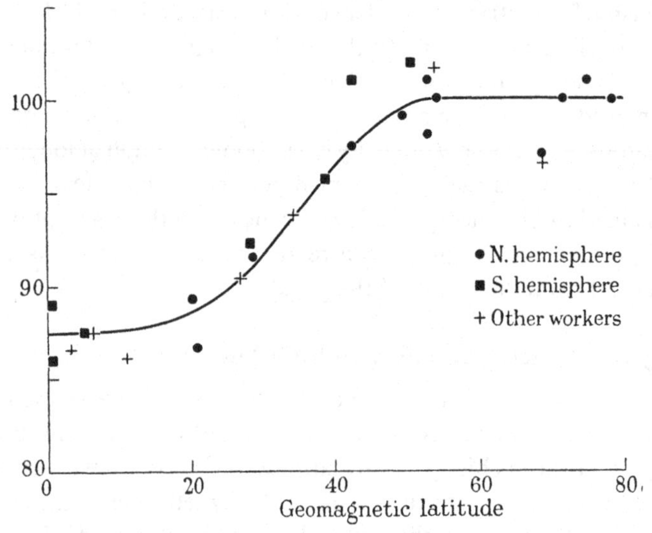

Fig. 3·3. Sea-level results of Compton's World Survey (2).

the magnetic moment of the sun, as deduced from the Zeeman effect in the spectrum of the sun's limb, is sufficient to account for the prohibition of particles of energy less than $3 \cdot 10^9$ e.v. If the "solar effect" is real we should expect a diurnal variation of cosmic-ray intensity due to rotation of the earth with respect to the solar cones of partial access, but the magnitude of this variation is probably rather too small to allow the detection of the effect in the existing experimental data.

REFERENCES

(1) Alfvén, *Phys. Rev.* **54**, 97, 1938.
(2) Compton, *Rev. Sci. Inst.* **7**, 71, 1936; *Phys. Rev.* **52**, 799, 1937.
(3) Rossi, *Phys. Rev.* **36**, 606, 1930; Lemaître and Vallarta, *Phys. Rev.* **43**, 87, 1933; Vallarta, *Phys. Rev.* **44**, 1, 1933.
(4) Johnson, *Phys. Rev.* **48**, 287, 1936.
(5) Lemaître, Vallarta and Brouckaert, *Phys. Rev.* **47**, 434, 1935; Vallarta, *Phys. Rev.* **47**, 647, 1935.
(6) Cosyns, *Nature*, **137**, 616, 1936.
(7) Carmichael and Dymond, *Nature*, **141**, 910, 1937.
(8) Jánossy, *Z. Physik*, **104**, 430, 1937.

For an extensive bibliography of the geomagnetic cosmic-ray effects see T. H. Johnson, *Rev. Mod. Phys.* **10**, 194, 1938.

Chapter IV

THE ABSORPTION OF THE RAYS IN THE ATMOSPHERE AND IN OTHER MATTER

4·1. The variation of cosmic-ray intensity with altitude.

The variation of cosmic-ray intensity with altitude has been investigated most fully with small unmanned balloons which

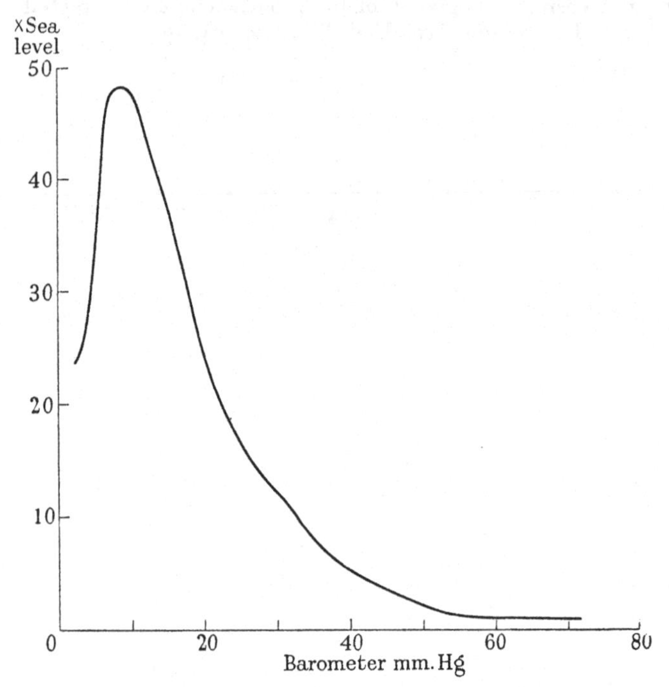

Fig. 4·1. Vertical intensity of rays. Variation with altitude (Pfotzer).

carry the apparatus to within 1–2 per cent of the top of the atmosphere. A typical curve (1) obtained with vertical coincidence counters in geomagnetic latitude 49° N. is given in Fig. 4·1, and its details will be discussed in connection with the theory of the absorption of the cosmic particles. A most conspicuous feature of the curve at high altitudes is the maximum at about 8 cm. Hg from the top of the atmosphere. This maximum is confirmed by ionisation-chamber measurements, and it indicates the production of a large secondary radiation in the top part of the atmosphere. The small inflection at about 30 cm. Hg is regarded by Pfotzer as experimentally established, but its interpretation is not yet clear.

4·2. Balloon experiments have now been made in a number of different latitudes (2). Since in each latitude the earth's magnetic field sets a lower limit to the energy spectrum of the incident particles, a comparison of the results at different latitudes enables us to deduce:

(1) The energy spectrum of the incident rays.

(2) The behaviour in the atmosphere of particles confined to a definite energy range.

Fig. 4·2 gives the results obtained in four different latitudes by Bowen, Millikan and Neher, using an ionisation chamber. The ordinates here represent ion production per sec. per c.c. of air at N.T.P., and the energy dissipated per c.c. per sec. may be obtained by multiplying the ionisation by the energy required to produce an ion pair (taken as 32 e.v. by B., M., N.). The energy carried out of the atmosphere into the ground by the rays at sea level is small compared with that dissipated in the atmosphere, and the energy dissipated in a column of the atmosphere of unit cross-section may be obtained by integrating the curve connecting depth and ionisation energy per c.c. The area under one of the curves of Fig. 4·2 is therefore proportional to the energy brought into 1 sq. cm. at the top of the atmosphere by the incident rays in the appropriate latitude. The ionisation chambers used in the measurements were sensitive to rays coming in any direction, but as a simplifying approximation the energy spectrum of the rays entering the atmosphere at any latitude is supposed to be cut off at the energy corresponding to the limit for vertical incidence, as calculated by Lemaître and Vallarta. The difference between curves A and B is, then, due to the rays having energies between

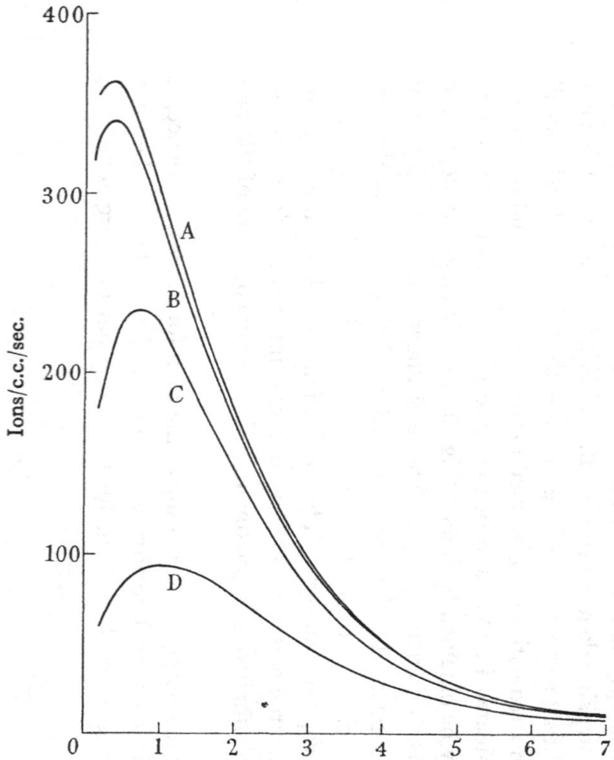

Fig. 4·2. Balloon experiments of Bowen, Millikan and

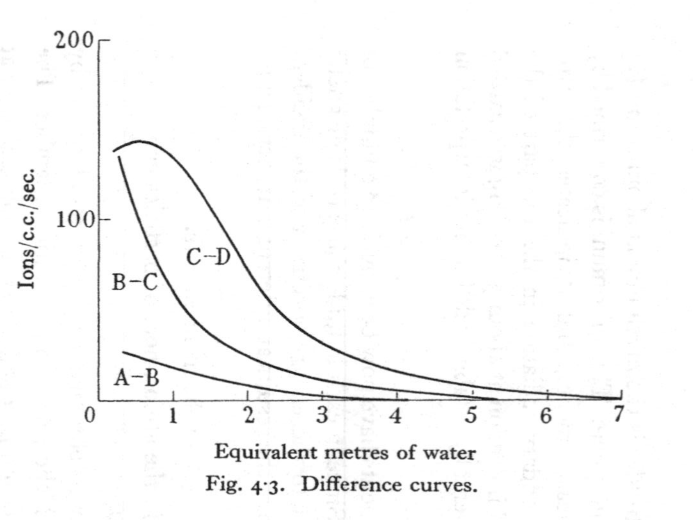

Fig. 4·3. Difference curves.

Results of balloon flights at different latitudes

	Place	Geo-magnetic latitude	Minimum energy vertical	Mean energy	
A	Saskatoon, Canada	60° N.	$1 \cdot 4 . 10^9$ e.v.	$2 \cdot 1 . 10^9$ e.v.	*A–B*
B	Omaha, Neb.	51° N.	2·9	4·8	*B–C*
C	San Antonio, Tex.	38° N.	6·7	10	*C–D*
D	Madras, India	3° N.	17		

$1 \cdot 4 . 10^9$ e.v. and $2 \cdot 9 . 10^9$ e.v. These rays have average energy about $2 \cdot 1 . 10^9$ e.v., and the number of incident rays in this energy range may be obtained by dividing the total energy (the area of the difference curve *A–B*) by this average energy. This operation, repeated with the curves *B–C*, *C–D*, gives the energy spectrum of the incident rays, which may be represented by the diagram, Fig. 4·4. The areas of the blocks here represent the numbers of incident particles in the various energy groups. The area of the fourth block is obtained by dividing the total energy at the equator

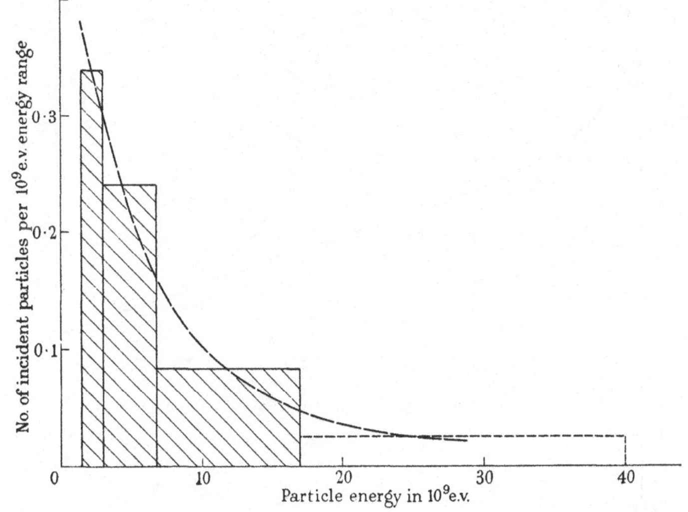

Fig. 4·4. Incident energy spectrum.

(curve *D*) by an assumed average particle energy of $20 . 10^9$ e.v. It may include an unknown contribution due to incident photons (of any energy). Bowen, Millikan and Neher calculate that the energy brought into the atmosphere by photons is less than 20 per cent of that brought by charged particles. There is no evidence about the *number* of incident photons or their individual energy.

The behaviour of electrons of energy lying in a definite range, as they pass through the atmosphere, may be deduced directly from the difference curves of Fig. 4·3. It is compared with theoretical considerations in § 6·8.

4·3. The absorption of the rays.

An absorption curve for the cosmic rays may be obtained by putting layers of absorbing material between the counters of a vertical coincidence counter system (3, 4). Such a curve shows an initial fall, followed by a very slow decrease (Fig. 4·5). The radia-

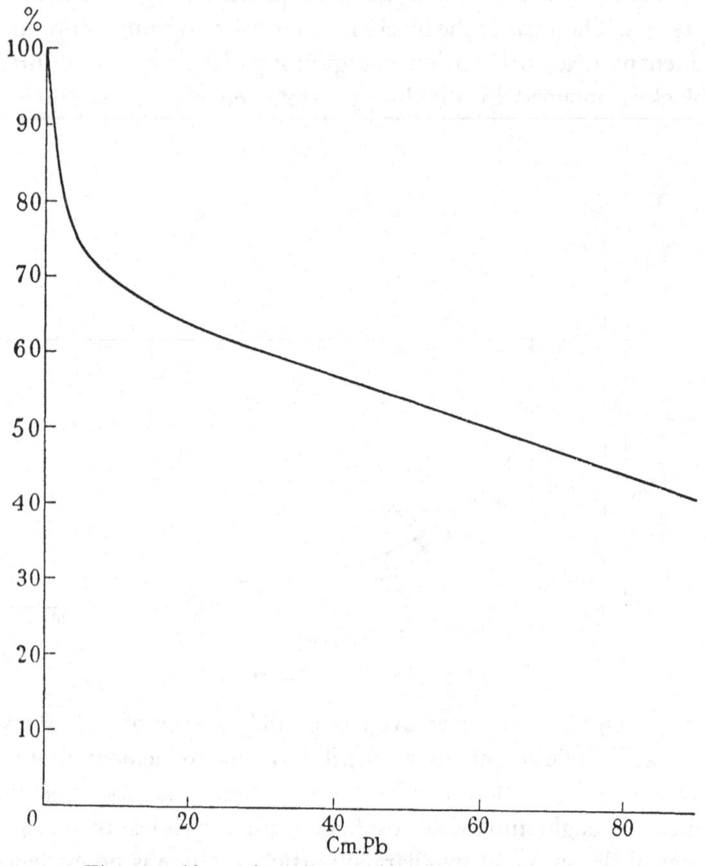

Fig. 4·5. Absorption of vertical rays in lead at sea level.

tion may therefore be divided, at least roughly, into a hard and a soft component. At sea level, about 20 per cent of the rays are absorbed by 10 cm. of lead, and the absorption beyond this thickness is practically that characteristic of the hard component. The absorption of the hard radiation depends very nearly on the mass per unit area of the absorbing layer; the absorption curves for different substances, plotted against mass per unit of absorber, are coincident in this region. Since the number of electrons per atom is nearly proportional to the atomic mass, this law of absorption indicates that the absorption per electron is constant.

The measurement of the absorption of the soft component is complicated by the production of secondary rays in the first layers of absorber (transition phenomena), and the results obtained depend on the precise recording arrangement, since different arrangements record different proportions of the secondaries.

Absorption measurements taken with an ionisation chamber, which, in effect, records every ray even if several pass simultaneously, show an increase in the radiation when a thin layer of heavy material is interposed (5). This shows the production of secondaries in excess of the number of particles absorbed, and is in accord with the experiments on shower production. In a counter system, on the other hand, several coherent secondaries give rise to only a single counter discharge, and the ratio

$$\frac{\text{counting rate with absorber}}{\text{counting rate without absorber}}$$

is the probability that at least one particle shall emerge from the layer for each particle incident.

This probability has recently been calculated with the aid of the quantum theory of showers (6), which, however, treats the production of secondaries as one-dimensional. A further complication may in practice be introduced by the production of secondaries making an angle with the primary ray, and so increasing the number of counts when the primary itself would have missed the lower counter. The chance of this depends on the geometry of the recording arrangements.

Any measurement of the absorption of these soft rays is, then, a measurement of a quantity arbitrarily defined by the measuring arrangement. *Comparative* measurements with a particular counter system led Auger to the conclusion that the absorption of the soft rays, per atom of absorbing element, increased more rapidly than Z, the atomic number.

4·4. Absorption measurements at high altitude.

Absorption measurements at high altitude show that the proportion of soft radiation is higher than at sea level (Table 4·1 (4)).

TABLE 4·1

Altitude	Barometer	Per cent soft
Sea level	760 mm.	20
3500 m.	510	48
9 km.	220	58

The absorption properties of the two types of radiation do not vary very much with height in the region considered here, but the penetrating component becomes more penetrating on filtration by thick layers of material.

4·5. The cosmic rays at great depths.

The penetrating component of the cosmic rays is responsible for an appreciable cosmic-ray intensity under very thick layers of absorbing matter, and Fig. 4·6 shows the results of measurements made with a set of vertical coincidence counters lowered into deep water to a depth of 240 m. (7). Fig. 4·6 includes also several points determined under layers of other materials and reduced to equivalent depth of water on the assumption of equal absorption in equal mass of absorber.

The absorption curve for air may be obtained by measurements of rays incident at different angles. The rays are assumed initially isotropic, and an air layer of variable thickness is interposed. It shows a larger absorption for equal mass. These measurements and their interpretation will be discussed in § 7·5.

There is evidence that the cosmic rays at a considerable depth below sea level still contain a considerable proportion of soft radiation. If sheets of heavy metal, e.g. lead, are placed between the vertical counters in an underground or underwater measurement, the absorption curve shows a steep initial fall as at sea level. Neither the experimental facts nor their interpretation are quite clear. If the secondary soft rays are in equilibrium with hard primaries, the absorption in further sheets of absorber in which the equilibrium proportion of primary and secondary is the same as in the incident rays might be expected merely to follow the absorption curve of the hard component. The observation of a

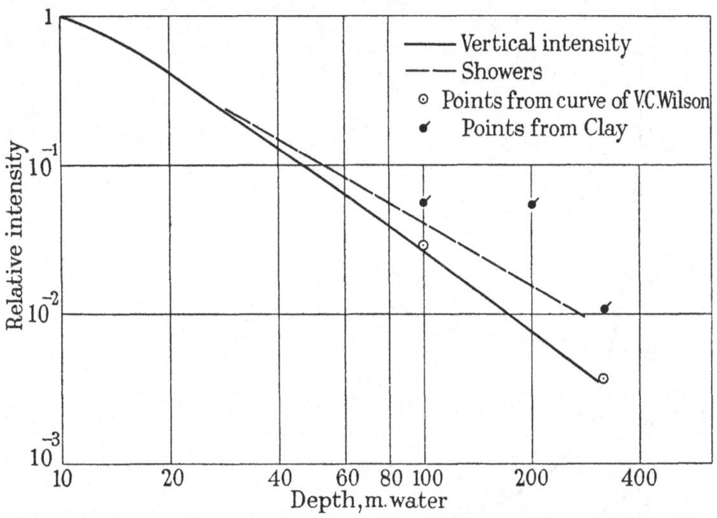

Fig. 4·6. Absorption of vertical rays and showers in deep water.

transition effect may possibly be explained as follows. In the rays incident on the top of the counter system, primary and secondary rays are not in general simultaneous, since the secondaries have originated at distances large compared with the size of the counter and the secondaries have some angular divergence (incoherent secondaries). The secondaries emerging from the bottom of the absorbing block, which is of high density and very close to the counter, are more often simultaneous with primary rays and hence are not counted. Hence a smaller proportion of the emergent than of the incident secondaries are counted as separate rays, and there is an apparent absorption of the soft component.

The quantitative interpretation of these absorption experiments has not yet been made, but the presence of soft particles seems certain.

The absorption coefficient of the "hard" rays, measured under thick absorbing layers, is considerably smaller than that measured at sea level, and Fig. 4·7 shows the absorption coefficient at various depths, calculated from V. C. Wilson's measurements of vertical coincidences in a deep mine (8).

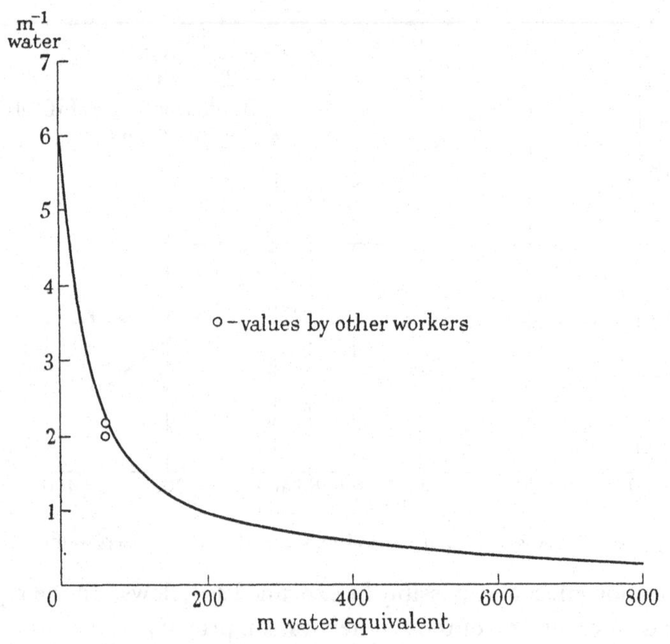

Fig. 4·7. Absorption coefficient of cosmic rays measured underground.

It is likely that this hardening of the penetrating radiation by filtration is due to a change in its energy distribution.

4·6. The energy spectrum of the rays.

The frequency of occurrence of rays as a function of energy has been studied, particularly by Blackett, by the use of a counter-controlled cloud chamber in a magnetic field (9). It was found that the particles at sea level were very nearly equally divided between positive and negative signs of charge. In the energy region between $3 . 10^9$ e.v. and the upper limit of measurement at $2 . 10^{10}$ e.v. the distribution could be expressed approximately by a relation

$$F(E) = AE^{-2},$$

where $F(E)\,dE$ is the number of particles with energies between E and $E + dE$.

There seems to be very little good direct experimental evidence on the energy spectrum of the particles in the energy range below 10^9 e.v. It is now known that the position is complicated by the existence of two kinds of particles, the "hard" rays and the "soft" electronic rays, and several attempts have been made to calculate the energy spectrum of the soft rays indirectly (see § 6·7). More direct experiment on this point is needed.

REFERENCES

(1) Pfotzer, *Z. Physik*, **102**, 23, 1936.
(2) Bowen, Millikan and Neher, *Phys. Rev.* **53**, 855, 1938.
(3) Rossi, *Z. Physik*, **82**, 151, 1933; Auger and Rosenberg, *Jour. de Phys.* **6**, 229, 1935.
(4) Auger, Leprince-Ringuet and Ehrenfest, *Jour. de Phys.* **7**, 58, 1936.
(5) Schindler, *Z. Physik*, **72**, 625, 1931.
(6) Arley, *Proc. Roy. Soc.* **168**, 519, 1938.
(7) Ehmert, *Z. Physik*, **106**, 751, 1937.
(8) Wilson, *Phys. Rev.* **53**, 337, 1938.
(9) Blackett, *Proc. Roy. Soc.* **159**, 1, 1937.

Chapter V

THE EXPERIMENTAL CHARACTERISTICS OF SHOWERS AND BURSTS

5·1. The characteristics of showers.

The occurrence of showers of simultaneous rays is an important phenomenon connected with the cosmic radiation. If a coincidence counter system is set up in such a way that more than one particle is required to produce a coincidence (Fig. 2·1 *b*), the rate of counting is small if there is no heavy material in the neighbourhood. The rate is increased by putting a layer of heavy matter, e.g. metal, above the counters, so that showers are produced by the passage of the rays through matter. If successive layers of shower-producing material are put above the counter system, we obtain a curve (the Rossi transition curve) showing how the number of showers varies with the thickness of material. It must be remembered that every counter system responds only when at least one ray goes through each counter, and every arrangement therefore selects showers according to an arbitrary criterion.

The shape of the transition curves depends to some extent on the counting arrangement used, but the main features are quite well defined. In the first part of the curve, all the workers have obtained a rapid initial increase of shower frequency with thickness of material, a maximum count under 1–2 cm. of lead, and a rapid fall at greater thicknesses (Fig. 5·1, curve 1). The thickness required for the maximum depends to some extent on the counter arrangement, and there is some evidence that the maximum occurs at a greater thickness when the counter system requires three or more coincident rays than in the cases where two rays are sufficient (7).

When different elements are used for the production of showers, the number of showers per atom of shower-producing material increases with increasing atomic number. Since the shower transition curve does not give a linear increase of showers with thickness it is not possible to define a "shower-production cross-section" which could be used to compare different substances, but it has been found that the shower production in thin layers

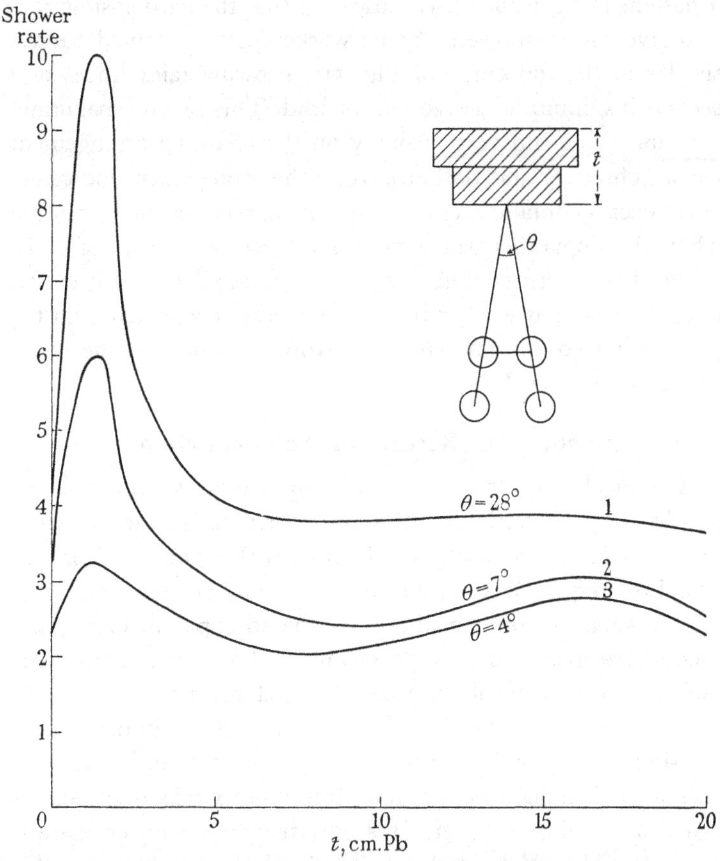

Fig. 5·1. Shower transition curves (Schmeiser and Bothe).

containing the same number of atoms varies approximately with Z^2.

5·2. The rapid decrease of the showers after the maximum indicates that the radiation producing the greater part of the showers is rapidly absorbed. We shall see that it is probably identifiable with the soft component observed in absorption measurements and consists of a mixture of electrons and photons.

If the transition curve is investigated up to greater thicknesses, it flattens out considerably, indicating that the hard cosmic rays also give rise to showers. Some workers have obtained curves, similar to the top curve of Fig. 5·1, showing faint traces of a second maximum at 15–20 cm. of lead. This second maximum appears to depend very strongly on the counting arrangement used. Schmeiser and Bothe (1), using the arrangement indicated, have recently obtained a more strongly marked second maximum when the apparatus was arranged to record showers of small angle. The interpretation of this second maximum, if it exists, is obscure, but it is clear that some process connected with the penetrating cosmic-ray particles involves a range of the order 15–20 cm. of lead.*

5·3. The study of showers in the cloud chamber.

A typical shower, as revealed by cloud-chamber photographs (2, 3), is a fairly complex phenomenon. There may be a number of tracks crossing the chamber without obvious connection, but frequently a number of tracks may be seen to diverge from a small region, e.g. there may be an apparent origin in a metal plate placed inside the chamber. There may be a track incident on the plate above a shower which diverges from a point in such a plate, or the shower may appear to be initiated by a non-ionising agency. It is noteworthy that frequently there are several such radiant points, and often other tracks occur distributed about the chamber. It seems that in a shower there is a number of particles and non-ionising links which are capable of producing further showers. On purely experimental grounds, therefore, it appears that many large showers are produced by a cascade process. The process is particularly well illustrated by chamber experiments with several thin horizontal lead plates in the chamber. The multiplication of the shower particles, as

* Data now available (June 1939) make the existence of the second maximum very doubtful.

they go through a plate, and the absorption of the particles in subsequent plates are then clearly visible. The quantum theory shows that this behaviour is to be expected of fast electrons. These produce radiation in passing through matter, and the radiation produces pairs of electrons and leads to a cascade. This theory and the experiments directly suggested by it will be discussed in §6·6.

The cloud-chamber experiments provide occasional rare examples of a different kind of shower, which may be designated as "explosive". The particles in an explosive shower seem to come from a single point, and are distributed in all directions. In the ordinary showers, the particles are usually more or less parallel. In explosive showers there are particles which ionise more heavily than electrons—very probably these include protons, but it is possible that they are slow, heavy electrons (§7·3).

Explosive showers are very rare compared with showers of the ordinary type; in Fussell's (3) investigation with a counter-controlled chamber, three photographs out of nine hundred showed showers of this type.

5·4. The production of large showers in the atmosphere.

Recent experiments by Auger and his collaborators (4) have shown that coincidences can be observed with Geiger counters separated by horizontal distances up to 75 m. These coincidences are due to very large showers produced in the air. It was shown that in these showers the density of particles might be of the order of 50 per sq. m., and that the greater number of the particles were electrons. The interpretation of these showers is discussed in §6·8.

5·5. The bursts or Stösse.

The lower limit of observation of bursts in an ionisation chamber is set by statistical fluctuations (§ 2·3). This limit is in practice about 10^5 ions, corresponding to the simultaneous passage of 10 electronic rays. Bursts of 10^9 ions have been observed, and if these were entirely electronic they would correspond to the passage of about 100,000 rays. The frequency of bursts decreases as their size increases, and the empirical size-distribution curve of the bursts in a given chamber fits roughly to the formula (5),

$$RdN = \frac{A}{N^s} dN,$$

where RdN = rate of occurrence of bursts with number of rays between N and $N+dN$;

$$s \sim 3\text{–}4.$$

The smaller bursts, then, correspond to showers of a kind frequently observed in cloud chambers, but on account of their rarity there is little chance of examining large bursts in this way.

It is necessary to ask if these bursts are of the same nature as the showers. Evidence that they are mainly identical is provided by the following facts:

(1) A few cloud-chamber photographs exist which show typical electronic showers of sufficient size to give bursts of several hundreds of rays.

(2) The size-distribution curve of the bursts is smooth and monotonic over a large range.* The smallest bursts are certainly identical with electronic showers, and this continuous relation suggests that the large bursts are of similar nature.

(3) Experiments on the efficiency of collection of the bursts in an ionisation chamber and its variation with gas pressure indicate that most of the particles are thinly ionising.

(4) Simultaneous records of bursts and coincidence-counter discharges show that many of the counts are accompanied by bursts (6), and that the number of occurrences observed by one arrangement and missed by the other are in accordance with statistical expectation.

It has been shown that the cascade theory is adequate to provide explanation of quite large bursts (§6·7). It is, of course, possible that a fairly small proportion of both bursts and showers is not of this type, and indeed there is direct evidence of occasional "explosive" showers (§5·3).

* Some of the earlier workers found a maximum in the curve and concluded that showers and bursts were distinct phenomena.

5·6 The variation of showers and bursts with altitude.

The showers and bursts increase with altitude more rapidly than the vertical cosmic-ray intensity or the intensity measured by an ionisation chamber (Table 5·1).

TABLE 5·1

Increase of bursts and showers with altitude

(*a*) Showers as recorded by coincidence counters

Reference	Altitude	Barometer	Ratio of intensity to sea-level value		
			Showers	Vertical	"Soft" vertical
	km.	cm.			
Auger, etc., *Jour. de Phys.* **7**, 58, 1936	3·5	49·5	6?	2·5	4·5
Woodward and Street, *Phys. Rev.* **49**, 198, 1936	3·1	52	—	1·9	3·2
Woodward, *Phys. Rev.* **49**, 711, 1936	3·2	51	5·0	2·5	—
	4·3	44	8·5	3·6	—
Braddick and Gilbert, *Proc. Roy. Soc.* **156**, 570, 1936	9	22	57	16	58 (rough estimate)

(*b*) Bursts of different sizes, measured in ionisation chamber under 1·3 cm. Pb (R. T. Young, *Phys. Rev.* **52**, 559, 1937)

Altitude	Barometer	Size of burst	Ratio of frequency to sea level
km.	cm.	rays	
3·2	51	10–19	5·1
		20–29	6·6
		> 30	7·1
4·3	44	10–19	8·1
		20–29	14·8
		> 30	14·1

(*c*) Showers of different sizes, measured in a cloud chamber (Anderson and Neddermeyer, *Phys. Rev.* **50**, 263, 1936)

Altitude	Barometer	Size of shower	Ratio of frequency to sea level
km.	cm.	tracks	
4·3	44	2–4	8·6
		5–10	21
		11–100	29

The shower intensity measured by a counter set increases roughly proportionally to the soft component of the rays. The increase in frequency of bursts of considerable size (30–100 rays) is still more rapid, at least in the lower part of the atmosphere.

Showers have been observed in experiments under water and in underground workings, and cloud-chamber photographs show that most of the showers at least are of the ordinary electronic type. The curve of shower intensity against depth shows that the shower frequency does not decrease more rapidly with depth than the general radiation (as might perhaps be expected from the behaviour of showers above sea level). In fact it appears that the ratio of showers to vertical intensity increases slowly with increasing depth (Fig. 4·6).

These facts suggest that the shower-producing rays are probably identical with the soft component of the rays. Above the earth's surface, the proportion of shower-producing rays to total rays is an increasing function of altitude (cf. § 4·4), below sea level the soft and hard rays are probably in equilibrium and the slow increase in relative shower intensity may be correlated with the changing character of the hard radiation which is shown by its increasing hardness.

REFERENCES

(1) Schmeiser and Bothe, *Ann. der Phys.* **32**, 161, 1938.
(2) Blackett and Occhialini, *Proc. Roy. Soc.* **139**, 699, 1933.
(3) Fussell, *Phys. Rev.* **51**, 1005, 1936.
(4) Auger, Maze, Ehrenfest and Fréon, *Jour. de Phys.* **10**, 39, 1939.
(5) Montgomery and Montgomery, *Phys. Rev.* **48**, 786, 969, 1935.
(6) Ehrenberg, *Proc. Roy. Soc.* **155**, 532, 1936.
(7) For a bibliography of showers and bursts see Stearns and Froman, *Rev. Mod. Phys.* **10**, 133, 1938.

Chapter VI

THE APPLICATION OF THE THEORY OF ELECTRONS TO COSMIC RAYS

6·1. The theory of electrons.

The interpretation of cosmic-ray phenomena has been greatly clarified in the last two years by the application of the quantum theory to the passage of fast particles through matter, and it has become clear how far the observations can be accounted for by the presence of energetic electrons.

When an energetic electron passes through matter, it loses energy in two main ways:

(1) It loses energy by collision with the extranuclear electrons of the atoms through which it passes. Some of these electrons are removed from their atoms (primary ionisation) and some atoms are thrown into excited states.

(2) When the electron passes through the strong electric fields surrounding the atomic nucleus, it is subject to large accelerations and gives up some energy in radiation.

Both of these processes may be treated in an approximately quantitative manner by the relativistic quantum mechanics.

6·2. The collision loss.

The energy lost by collisions goes partly into direct "primary" ionisation and partly into excitation of atomic levels. The electrons liberated by primary ionisation, and the quanta resulting from excitation, in turn produce further ionisation ("secondary" ionisation) and excitation. On the average, the total energy dissipated for the formation of a single ion pair is considerably greater than the ionisation potential. Experimentally, the energy dissipated per ion pair seems to be fairly constant for all fast-moving ionising particles in a given medium. The best value for air, obtained from measurements on electrons of a few thousand electron-volts energy, is 32·2 e.v.

The quantum theory gives, for the total energy loss of a fairly fast electron, by collision, per cm. path (1),

$$-\frac{dE}{dx}=NZ\phi_0\frac{1}{\beta^2}\left\{\log\left\{\frac{(E-\mu)E^2\beta^2}{2\mu I^2Z^2}\right\}+\left(\frac{\mu}{E}\right)^2\right\}, \quad (6\cdot21)$$

where N = number of atoms per c.c.,

Z = atomic number,

μ = rest energy of electron = mc^2,

I = average ionisation energy of an electron in the atom. It may be found by experiments, e.g. on the stopping of α-particles,

ϕ_0 = unit cross-section

$$=\frac{8\pi}{3}\left(\frac{e^2}{mc^2}\right)^2=6\cdot57.10^{-25}\text{ cm.}^2,$$

E = total energy of particle $=\dfrac{mc^2}{\sqrt{1-\beta^2}}$.

The energy loss by collision, in the case of an electron, decreases

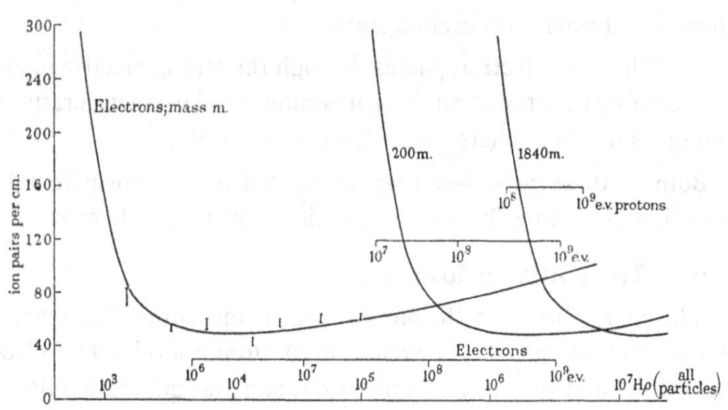

Fig. 6·1. Calculated ionisation by particles of mass m, $200m$, $1840m$. Experimental points for electron from Corson and Brode.

rapidly to a minimum at about $E=2mc^2$ (kinetic energy about $0\cdot5.10^6$ e.v.), and then rises slowly in the same general way as the primary ionisation plotted in Fig. 6·1. The collision energy loss of a particle going through matter, equation (6·21), depends approximately on the total number of electrons per c.c., changes in I from element to element being comparatively small. Since the atomic masses and the atomic numbers of the elements are closely proportional, the energy loss depends nearly on the mass of absorber traversed.

The theory shows that a heavier particle with unit charge loses practically the same energy per cm. path as an electron of the same *velocity*. It therefore loses more energy than an electron of the same energy, except at extreme relativistic energies ($>Mc^2$, where M is the mass of the particle).

6·3. Ionisation by collision.

The theoretical variation of the primary ionisation with velocity is given by the expression

$$\frac{A}{\beta^2}\left\{\log\frac{\beta}{1-\beta^2}+\log k-\beta^2\right\}, \qquad (6\cdot31)$$

which shows a rapid fall to a minimum and a slow logarithmic rise.

The "probable" ionisation, which may be directly measured by counting the drops in a cloud-chamber track, includes a part of the secondary ionisation, but it has been shown that it differs from formula (6·31) only in the value of the constant k.

Fig. 6·1 shows the ionisation found experimentally in this way by Corson and Brode (2), and a theoretical curve calculated with the constant k chosen as 2.10^4. It is clear that there is some experimental evidence for the existence of the theoretical rise at high energies.

In Fig. 6·1 there are also plotted the calculated densities of ionisation due to singly charged particles of mass $200m$ and $1840m$ (protons). Since a cloud-chamber track can be distinguished by simple inspection if the density of ionisation is about four times that of an ordinary high energy electron track, it is clear that a proton track is distinguishable by inspection below about 10^8 e.v. energy ($H\rho$ about 10^6), and a particle of mass $200m$ is distinguishable below 10^7 e.v. ($H\rho$ about 10^5).

6·4. Radiation by a fast electron.

Radiation quanta are emitted when the moving electron undergoes acceleration in the electric field surrounding an atomic nucleus. The nuclear field is modified by the screening effect of the atomic electrons, and the screening is greatest when the moving electron has a high energy, because high energy electrons produce effects at large distances from the nucleus.

The rate of loss of energy, for high energy electrons (case of complete screening) is given by

$$-\frac{dE}{dx}=\frac{NZ^2}{137}\left(\frac{e^2}{mc^2}\right)^2 E\left(4\log 183Z^{-\frac{1}{3}}+\frac{2}{9}\right). \quad (6\cdot 41)$$

At high energies, then, the electron loses a constant fraction of its energy E by radiation in each centimetre of its path.

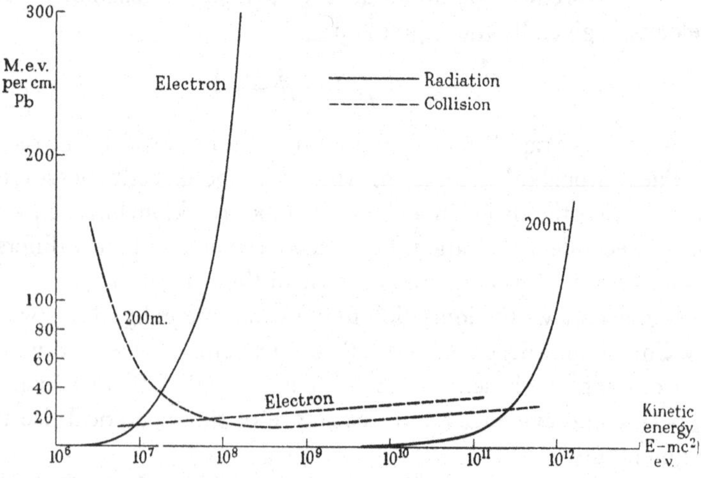

Fig. 6·2. Radiation loss in lead for particles of mass m, $200m$.

The radiation loss in a given element increases with the square of the atomic number, and energy loss by radiation is more important, compared with collision losses, in the heavy elements.

Fig. 6·2 shows the radiation and collision losses calculated for lead. It shows that for kinetic energies above $20mc^2$ (10^7 e.v.) the radiation loss becomes predominant. In water the comparable energy is about $250mc^2$ ($> 10^9$ e.v.).

If we consider a particle more massive than the electron, then, apart from small changes in the screening effect, the radiation is reduced in the ratio m^2/M^2, where m is the mass of the electron and M that of the particle.

Fig. 6·2 includes a radiation loss curve calculated for a particle of mass $200m$, and it may be seen that the radiation loss is unimportant compared with collision up to energies of over 10^{11} e.v. For a proton the radiation loss is negligible up to 10^{12} e.v.

The radiation loss as calculated is also characterised by a high probability that an electron will radiate a large fraction of its energy at a single encounter. This fact has two important results:

(1) There is a large "straggling" of the particles, and different particles of the same energy may experience very different energy losses in going through a given thickness of absorber.

(2) The quanta produced may have very high energies, and they may give rise in turn to energetic secondary electrons. It is this property which gives the possibility of cascade-shower production (§ 6·6).

6·5. Direct comparison of energy loss with experiment.

Using electrons from radioactive sources, experiments have been made on the energy loss of electrons of energies up to about $20mc^2$ (11 M.e.v.) in lead plates. The agreement with theory is only fair, the average experimental energy loss being higher than the theoretical, and the frequency of large energy losses (of the order of half the initial kinetic energy) very much higher than the theoretical (3).

In the region of cosmic-ray energies, cloud-chamber measurements have shown the existence of particles whose energy loss is given fairly accurately by the theory. According to Blackett (4) practically all the particles with energies below $2 \,.\, 10^8$ e.v., when divided into energy groups, show an average energy loss in each group comparable with the theoretical value, and the distribution of the energy losses of individual particles is in accordance with the statistical predictions of theory. A few particles of energies greater than $2 \,.\, 10^8$ e.v. show the expected energy loss for electrons, but most of the particles in this upper range show a much smaller energy loss.

The experiments of Neddermeyer and Anderson (5) also show

particles of two kinds, but they find both electrons having energy losses compatible with the theory, and less absorbable particles, throughout the energy range 1·2–5 . 10⁸ e.v., and they observe that the particles showing electronic behaviour are usually associated with other particles in showers. The question of the less absorbable particles will be taken up in chap. VII. The experiments of Blackett and of Neddermeyer and Anderson agree in showing that the radiation formula correctly describes the behaviour of electrons up to about $5 . 10^8$ e.v. There is indirect evidence that it is valid at much higher energies (§§ 6·7, 6·8).

6·6. Application of the theory to showers.

The results given above indicate that when a fast electron (energy $\gg mc^2$) passes through matter there is a considerable loss of energy by radiation and a high probability that quanta of large energy will be radiated. Quanta of high energy ($\gg mc^2$) have, again, a large probability of producing positron-electron pairs in the strong fields near atomic nuclei (Table 6·1).

The theory therefore predicts the formation of secondaries and showers according to the cascade scheme of Fig. 6·3.

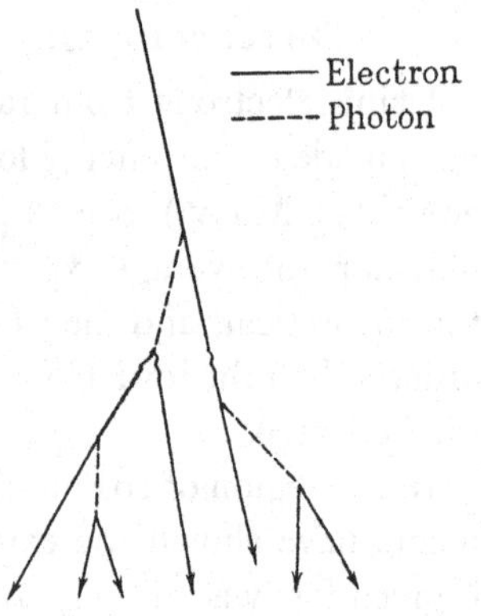

Fig. 6·3. Production of a cascade shower.

As the process of subdivision goes on, and the particles and quanta become less energetic, the system loses energy at an increasing rate by collision of the electrons and Compton effect of the quanta. The shower therefore attains a maximum size and is then absorbed, so that the initial rise and fall of the "shower transition curve" receive a qualitative explanation.

A shower of this type may, of course, begin with a photon of high energy instead of an electron, and the existence of photon links accounts for the production of apparently isolated subsidiary showers which occur in cloud-chamber photographs of showers.

At high energies, the quanta radiated by electrons and the paired electrons produced by quanta will be projected in the general direction of the original electron, positron or quantum. The mean angle which the secondaries make with the direction of the primary is of the order mc^2/E, where E is the energy of the secondary produced. As an approximation in the treatment of showers, the angular spread may be neglected and the problem treated as one-dimensional.

TABLE 6·1

Pair production in lead by quanta of energy hν

Energy $h\nu$	5	10	100	∞	mc^2
	2·5 . 10⁶	5 . 10⁶	5 . 10⁷	∞	e.v.
Atomic cross-section	2·7	7·5	30·4	44·2	10^{-24} cm.²
Mean free path of quantum	12	4·1	1·02	0·7	cm.

The quantitative treatment of the cascade-shower process was first carried out by Bhabha and Heitler and by Carlson and Oppenheimer (6). It has recently been refined by other workers.

The quantum theory gives for the probability per unit path that an electron or positron of energy E will radiate a quantum of energy between Q and $Q+dQ$ (1):

$$A\left\{(\log 183Z^{-\frac{1}{3}})\left(\frac{E^2+(E-Q)^2-\frac{2}{3}E(E-Q)}{E^2Q}\right)+\frac{2}{9}\frac{E-Q}{E}\right\}dQ, \quad (6\cdot61)$$

where $$A=\frac{4}{137}\left(\frac{e^2}{mc^2}\right)^2 Z^2\sigma,$$

σ = number of nuclei per c.c.

This may be put into the form

$$\frac{A'}{Q}f\left(\frac{Q}{E}\right)dQ, \quad A'=A\log 183Z^{-\frac{1}{3}},$$

where $f\left(\frac{Q}{E}\right)$ varies rather slowly with $\left(\frac{Q}{E}\right)$.

To a first approximation we may take the probability as

$$\frac{A'}{Q}dQ. \tag{6·62}$$

The corresponding formula for the probability of production of a pair with energy E, $Q-E$ by a quantum of energy Q is

$$A\left\{(\log 183Z^{-\frac{1}{3}})\frac{E^2+(Q-E)^2+\frac{2}{3}E(Q-E)}{Q^3}-\frac{2}{9}\frac{(Q-E)E}{Q^3}\right\}, \tag{6·63}$$

which gives, to a similar approximation,

$$\frac{A'\sigma}{Q}dE, \quad \sigma \simeq \tfrac{2}{3}. \tag{6·64}$$

Since $1/A'$ appears as a parameter in both (6·62) and (6·64) it is possible to choose $l = 1/A$ as a unit of length, and to obtain results for the multiplication process which are true for all substances.

Table 6·2 gives the calculated values of l for some important substances:

TABLE 6·2

Pb	H_2O	Air N.T.P.
0·4 cm.	35 cm.	275 m.

The calculation of the number of particles may be carried out step by step, using the formulae above for the number of γ-rays radiated by an electron and for the probability of pair production. Very roughly it appears that the number of rays doubles itself in the thickness l.

In order to allow for the absorption of the particles by collision processes, it is possible to calculate at each step the number of particles above the critical energy E_c at which the collision loss becomes important compared with radiation, and to reject all particles with energy less than E_c. Particles with energy less than E_c are rapidly absorbed by collisions. This method is that used by Bhabha and Heitler.

Alternatively, the collision loss may be assumed constant, and a diffusion equation set up to give the distribution in energy of the particles at any depth below the top of the absorber. This method was used by Carlson and Oppenheimer, and it has recently been found possible to make the calculations, using the exact formulae for radiation and pair production.

The general nature of the multiplication process may be shown by Fig. 6·4, which represents the results of Bhabha and Heitler.

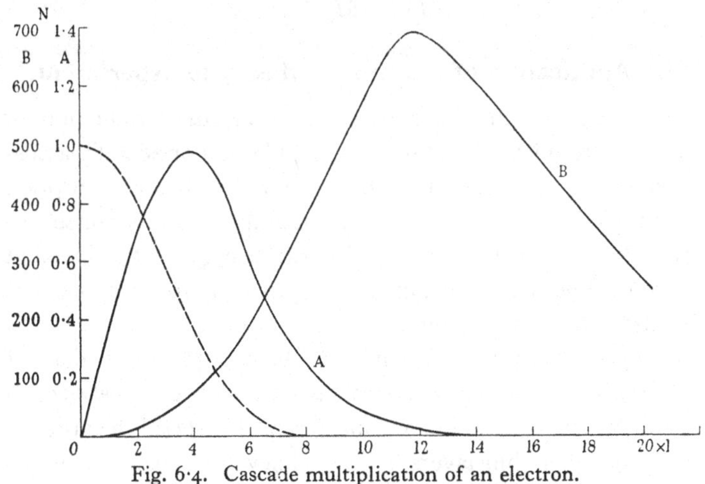

Fig. 6·4. Cascade multiplication of an electron.

Secondary production according to Bhabha and Heitler.

Curve A: $\log E/E_0 = 3$, $E = 2.10^8$ e.v. for lead.

Dotted curve: Absorption of primary electrons for $\log E/E_0 = 3$.

Curve B: $\log E/E_0 = 10$, $E = 2.10^{10}$ e.v. for lead.

Energy loss by collision has not been considered, and the ordinates represent the number of emergent electrons and positrons with energy greater than E_0.

In using these results, Bhabha and Heitler put $E_0 = E_c$, the "critical energy", which is 10^7 e.v. for passage through lead, and $1·5.10^8$ e.v. for air or water.

The multiplication is more rapid with incident particles of higher energy, and the thickness required to give maximum multiplication also increases with the energy of the incident particle. Since only a few encounters are needed for the production of a sizeable shower, there are large statistical fluctuations in the numbers of particles produced by electrons of given energy in a given thickness of material.

It has been calculated (7) that if the average number of rays produced in a single shower in a thin layer of material is M, the probability of finding as a fluctuation a shower of N rays is

$$\frac{1}{M}\left(1-\frac{1}{M}\right)^{N-1}.$$

6·7. Application of the shower theory to experiment.

Some experiments have been made on the production of showers in thin lead plates (0–1·6 cm.) placed horizontally across an expansion chamber. For thin layers of lead the absorption of the rays by collision is unimportant and the number of secondaries does not vary rapidly with the energy of the incident electrons. The multiplication observed was in rough agreement with theoretical prediction.

Montgomery and Montgomery (8) have applied the data of Fig. 6·4, and the fluctuation calculation referred to above, to account for the production of considerable bursts (of the order of 100 rays) under thin layers of lead. They find that the bursts can be produced by electrons with energies in the energy range 10^9–10^{12} e.v., and that the observed frequency distribution for bursts of different sizes is consistent with an incident energy distribution of the form

$$F(E) = AE^{2\cdot6},$$

where $F(E)\,dE$ is the frequency of electrons in the energy range between E and $E+dE$.

Further, the total number of electrons required in the energy range 10^9–10^{10} e.v. is only of the order 0·5 per cent of the observed number of particles in that part of the energy spectrum, and though it will be shown that most of the particles of high energy observed at sea level are not electrons, there is no difficulty in assuming enough electrons to account for the bursts. While a few bursts may be of a different type, these calculations make it probable that most bursts can be accepted as cascades.

6·8. The production of secondaries in the upper atmosphere.

The rapid increase in the number of cosmic rays with depth in the upper layers of the atmosphere can be explained in general terms by the cascade theory (§4·1). A special analysis of this problem has been made, using the *exact* quantum theory formulae for radiation and pair production, and an assumed constant value for collision loss (9).

Fig. 6·5 gives the curve calculated in this way for particles of 11.10^9 e.v. and modified to allow for isotropic incidence. The

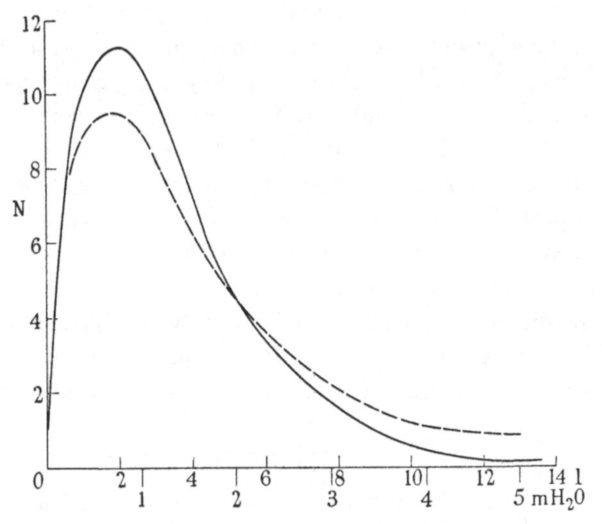

Fig. 6·5. Multiplication of electrons in the atmosphere.

dotted curve of Fig. 6·5 is the difference curve of Bowen, Millikan and Neher for latitudes 3° and 38°. It will be seen that the maximum in intensity occurs at about the right depth, but the theoretical maximum is rather greater than is observed experimentally. At a considerable depth, the theoretical number of particles is far lower than the observed. These differences are probably to be explained by the production and subsequent absorption of a penetrating secondary radiation.

Table 6·3 shows the average number of electrons which reach the bottom of the atmosphere for each electron of given energy incident on the top of the atmosphere. It is calculated from the formulae obtained by Serber on the assumptions detailed above:

TABLE 6·3

Initial energy in e.v.	10^{10}	10^{12}	10^{14}
Sea-level electrons per incident electron	$3 \cdot 10^{-4}$	3·5	5000

While the latitude-sensitive part of the incident electronic radiation produces a negligible direct effect at the bottom of the atmosphere, it will be seen that electrons with energy greater than 10^{12} e.v. produce cascades which extend down to sea level.

The large atmospheric showers of Auger (§ 5·4) are certainly parts of such cascades, and provide direct evidence for the existence of incident electrons of very high energy (10^{13}–10^{14} e.v.). The cascade electrons produced directly by electronic primaries form a part of the electron intensity (soft component) observed at sea level. Auger shows that the large showers contain 2–3 per cent of the total soft radiation, and a further part of the soft rays must be due to the straggling ends of showers which are mainly absorbed higher up. The remainder of the soft component requires the intervention of hard rays for its explanation (§ 7·9).

REFERENCES

(1) Bethe, *Handbuch der Physik*, XXIV (i); Heitler, *Quantum Theory of Radiation*. Oxford, 1936.
(2) Corson and Brode, *Phys. Rev.* **53**, 773, 1938.
(3) Turin and Crane, *Phys. Rev.* **52**, 63, 610, 1937; Laslett and Hurst, *Phys. Rev.* **52**, 1035, 1937.
(4) Blackett, *Proc. Roy. Soc.* **165**, 11, 1938.
(5) Neddermeyer and Anderson, *Phys. Rev.* **51**, 884, 1937.
(6) Bhabha and Heitler, *Proc. Roy. Soc.* **159**, 432, 1937; Carlson and Oppenheimer, *Phys. Rev.* **51**, 220, 1937.
(7) Furry, *Phys. Rev.* **52**, 569, 1937.
(8) Montgomery and Montgomery, *Phys. Rev.* **53**, 955, 1938.
(9) Serber, *Phys. Rev.* **54**, 317, 1938.

Chapter VII

THE PENETRATING COMPONENT OF THE RAYS

7·1. The theory of electrons and the penetrating rays.

It is clear that the quantum theory of radiating electrons explains several features of the cosmic radiation. It accounts for the multiplication of particles in the upper atmosphere, and for the production of showers. Its predictions agree with the properties of the soft radiation, but they do not explain how this soft radiation penetrates far enough into the atmosphere to be observed at sea level and to show latitude variation there. The absorption measurements show the existence of a hard component of the radiation, and it will appear that this cannot consist of electrons.

Experiments made with a cloud chamber under a considerable thickness of lead show that the hard rays often penetrate many centimetres of lead and emerge as single particles. An electron of any initial energy whatsoever, which behaved according to quantum theory, would nearly always be accompanied out of the lead by secondaries produced by the process discussed in chap. VI.

It would be possible to explain these results by postulating that the quantum theory of radiation breaks down at very high energies in such a way that the large radiation loss ceases and the electrons become penetrating.

This explanation ceases to be tenable if experiment shows that some particles of a given energy behave like radiating electrons and others of the same energy behave like penetrating particles. The following experimental evidence is available in this sense.

(1) The comparison of the difference curve of Bowen, Millikan and Neher with the cascade theory (§ 6·8, Fig. 6·5) shows that incoming particles in the energy range 8–17 . 10^9 e.v. multiply like radiating electrons in the upper atmosphere. The lower part of this curve shows, however, that particles in this energy range, or their

secondaries, can penetrate the whole thickness of the atmosphere. According to quantum theory the number of rays reaching the bottom of the atmosphere, due to incident electrons in this energy range, is much smaller than the number of particles observed (Fig. 6·5).

It may also be pointed out that incident particles of even lower energy must produce a considerable effect at the bottom of the atmosphere, since the latitude effect at sea level extends to $\lambda\,50°$. Auger has shown that the latitude effect is not diminished by the interposition of a 10 cm. lead screen, so that incoming particles of energy less than 10^{10} e.v. can give rise to penetrating particles at the bottom of the atmosphere.

(2) In the energy-loss measurements of Anderson, and of Blackett and Wilson, both penetrating and non-penetrating particles are found. The detailed results are still rather conflicting, since Anderson finds penetrating particles of low initial energy, while Blackett and Wilson find that particles below 2.10^8 e.v. all behave like electrons. But Blackett and Wilson find that in the energy range 2.10^8 e.v.–10^{10} e.v. there is a large number of penetrating particles, and that a small number of particles in the same energy range behaves like theoretical electrons.

Further evidence that electrons of high energy radiate more or less in accordance with theory is provided by the fact that bursts of considerable size are apparently produced by the cascade process, as shown by their properties forming a continuous sequence with those of recognised cascade showers. The total energy of such bursts must have belonged originally to a single electron of high energy.

Since the electron theory seems incapable of explaining the penetrating rays, the facts require the existence of particles whose energy loss by radiation is much less than that of electrons.

For particles of electronic charge having equal energy, the radiation loss is reduced by a factor m^2/M^2 when the mass is raised from m to M (§ 6·4), and the experimental results so far considered would be explained by assuming that the penetrating particles are protons. But protons of energy less than 5.10^8 e.v., which could be distinguished from electrons by their specific ionisation, are rare in chamber photographs, and Montgomery and Montgomery (1) estimate from the number of particles actually observed as protons, and the fraction of its range over which a proton can be distinguished as such, that only about 1 per cent of the hard cosmic rays could be protons.

7·2. The hypothesis of particles of intermediate mass.

It was first suggested by Anderson and Neddermeyer (2) that particles of mass intermediate between the proton and electron were present in the hard cosmic rays. Evidence on this point can be obtained by the analysis of Wilson-chamber photographs of the rays.

The specific ionisation of a particle of given velocity is independent of the mass of the particle and proportional to the square of its charge (§ 6·3). Its dependence on velocity is given by

$$I = \frac{A}{\beta^2}\left[\log k + \log\frac{\beta^2}{1-\beta^2} - \beta^2\right]. \qquad (6·31)$$

The fact that all *fast* particles observed in the cloud chamber have nearly the same specific ionisation indicates that all cosmic particles carry one unit charge.

Besides the curvature of the tracks in a magnetic field (expressed quantitatively by $H\rho$) and the specific ionisation, the following quantities may in some cases be obtained from cloud-chamber photographs:

(*a*) The energy loss of the particles in going through a metal plate, or in the gas which fills the chamber.

(*b*) The range of the particle.

For particles of mass M, with kinetic energies less than about Mc^2, the loss of energy is due almost entirely to collisions. For particles of the same velocity, but different masses, the range is proportional to the mass and the quantity $H\rho$ is proportional to the mass. The relation between range and $H\rho$ can therefore be obtained for particles of any assigned mass from the experimentally obtained relation for protons.

Corson and Brode (3) give a nomogram connecting mass, range, specific ionisation, $H\rho$, and change of $H\rho$ with distance (rate of energy loss in the gas). A knowledge of any two of these enables the others to be determined.

Table 7·1 gives all the observations of this kind yet published, and the masses deduced from them. It must be allowed that in many of these cases, the accuracy of the curvature measurements may be low on account of chamber distortions.

TABLE 7·1

Direct evidence for particles of intermediate mass

Investigators and reference	Notes on method	Quantities measured	M/m
Street and Stevenson, *Phys. Rev.* **52**, 1003, 1937	Special counter arrangement	I and $H\rho$	160
Nishina, Takeuchi and Ichimiya, *Phys. Rev.* **52**, 1198, 1937	Counter controlled. Energy loss in lead	$H\rho$ and ΔE	180–2[illegible]
Ruhlig and Crane, *Phys. Rev.* **53**, 266, 1938	Random expansion	$H\rho$ and $\frac{\Delta(H\rho)}{\Delta l}$	120
Corson and Brode, *Phys. Rev.* **53**, 773, 1938	Counter-controlled (*a*) delayed expansion (*b*)	$H\rho$ and I $H\rho$, range	250 $<$20[illegible]
Ehrenfest, *Comptes rend.* **206**, 428, 1938	Counter controlled	$H\rho$ and I	About [illegible]
Williams and Pickup, *Nature*, **141**, 684, 1938	Random expansion (*a*) (*b*) (*c*) (*d*)	$H\rho$ and I	220 ± [illegible] $>$430 $<$ [illegible] 190 ± [illegible] 160 ± [illegible]
Neddermeyer and Anderson, *Phys. Rev.* **54**, 88, 1938	Counter controlled with counter in chamber	Range and $H\rho$ I and $H\rho$	240
Maier-Leibnitz, *Naturwiss.* **26**, 677, 1938	Random expansions	$H\rho$, range	120 ± [illegible]

It is seen that there is considerable evidence for the existence of a particle of mass 150–250 times the electronic mass. The new particle has been called the "mesotron" and at present it seems possible to explain many of the phenomena of the hard cosmic rays by means of it. The investigation of its properties has been guided considerably by theoretical predictions.

7·3. Properties of particles of intermediate mass.

The existence of particles of mass 100–200 times the mass of the electron was postulated by Yukawa in 1935 to explain the binding forces inside the nucleus (4). Following Heisenberg's suggestion that the nuclear protons and neutrons were held together by "exchange forces", Fermi (in his theory of β-decay) considered the case where the interaction between proton and neutron involved the emission and absorption of an electron and neutrino. This view is unsatisfactory, because it gives nuclear forces which are far too small. Yukawa introduced a new type of field of force, associated with a new particle which is emitted and absorbed to produce the exchange forces. (According to quantum electrodynamics, the Coulomb forces between charged particles are similarly associated with the emission and absorption of photons.)

The forces between protons and neutrons are characterised by a rapid falling off at distances comparable with 10^{-12}–10^{-13} cm. Yukawa takes a potential of the form

$$\frac{Ae^{-\lambda r}}{r},$$

and shows that it is a solution of the wave equation

$$\nabla^2\phi - \frac{1}{c^2}\frac{\partial^2\phi}{\partial r^2} - \lambda^2\phi = 0,$$

and that the particle associated with the field has the rest-mass

$$m = \frac{\lambda h}{c}.$$

For $\lambda = 2.10^{13}$ cm., $m = 200m_0$, m_0 = electron mass.

Since the neutron and proton have half-integral spin and follow Fermi statistics the new particle must have integral spin and follow Bose statistics. The theory of the particles has now been worked out in some detail. The particles do not appear free in nuclear transformations, because the energy equivalent to their rest-mass (of the order 100 M.e.v.) is not available. The particles are of both signs (with one electronic charge) and a neutral particle of similar mass is required to explain, by exchange, proton-proton and neutron-neutron forces which have the same general character as the proton-neutron forces.

7·4. Spontaneous decay of the mesotron.

In order to account for the radioactive β-disintegration Yukawa suggested that the intermediate particles could interact with electrons and neutrons. In this interaction the mesotron with negative charge, produced virtually in the radioactive nucleus when a heavy particle jumps from a neutron to a proton state, is absorbed by a light particle which rises from a neutrino state of negative energy to an electron state of positive energy. This corresponds to the creation of an electron-antineutrino pair. It gives for the β-disintegration the same results as the theory of Fermi in which the proton-neutron transition is directly connected with the neutrino-electron transition, but since it separates the nuclear exchange forces from the β-decay it makes it possible to combine large nuclear forces with weak β-decay (5, 6, 7). Like the Fermi theory it involves some difficulty in explaining the shape of the β-ray energy distribution curves.

On account of this interaction, the mesotron has a certain probability of spontaneous decay into an electron-neutrino pair. The moving particle is a "clock" in the sense of relativity theory, and its average life, observed by a terrestrial observer, is lengthened as its velocity approaches the velocity of light, according to the formula $\tau = \frac{\tau_0}{\sqrt{1-\beta^2}}$. The average life and the mean free path for a particle of mass $200m_0$ has been calculated by Yukawa:—

TABLE 7·2

Kinetic energy	0	10^9	10^{10}	10^{11}	e.v.
Average life	$1{\cdot}3 . 10^{-7}$	$1{\cdot}3 . 10^{-6}$	$1{\cdot}3 . 10^{-5}$	$1{\cdot}3 . 10^{-4}$	sec.
Mean free path	—	0·4	4	40	km.

7·5. Experimental evidence for decaying particles in cosmic rays.

There is experimental evidence for a spontaneous decay of hard particles in the cosmic rays. It has been found that the absorption of the hard rays in air, obtained by inclining a set of counters at an angle θ to the zenith and thereby increasing the mass of air traversed in the ratio $1 : \sec\theta$, is greater than the absorption found by keeping the counters vertical and removing them to a station at lower altitude. The mass of air traversed is the same in each case, but the path length is greater in the case of the zenith-angle experiment, and the loss of particles by spontaneous disintegration is therefore greater.

It has also been found that the absorption of the hard rays for equal mass is less in heavy materials than in air, and this can clearly be explained if, in addition to an absorption dependent on mass, there is an independent spontaneous loss of particles increasing with the path traversed. Ehrenfest and Fréon(8) have compared the absorption of the atmosphere between two stations directly with the absorption of a layer of lead of equal mass, and they, as also Blackett and Rossi(9), have used the additional absorption to calculate the mean free path of the mesotron and its average life. The average life of the mesotron at rest comes out to be of the order $2 . 10^{-6}$ sec., and according to Blackett it is longer for the more than for the less penetrating hard particles, so that the variation of lifetime with velocity is approximately verified.

If the mesotrons have short lifetimes, they must be created close to the earth, and cannot be primary particles from outer space. An origin near the top of the atmosphere is assumed in the calculations of Blackett and of Rossi. The question of the origin of the hard rays will be dealt with in §7·6.

There is very little direct evidence from chamber photographs for the disintegration of the mesotron. There are indications of a decay electron in a photograph by Anderson and Neddermeyer, and in a photograph by Ehrenfest a particle which emerges from the wall of the chamber is probably a decay electron.*

7·6. The origin of the penetrating particles.

The idea of penetrating particles which decay spontaneously with a fairly short average life requires that the particles be created near the earth. Independent evidence for a secondary

* For references, see Table 7·1.

origin of the penetrating particles may be drawn from the following experimental results:

(1) The east-west asymmetry observed in the cosmic rays (§ 3·3) shows that a considerable majority of the incident particles are positively charged, at least in the energy range 10^{10}–$6 \cdot 10^{10}$ e.v. In observations in a cloud chamber in a magnetic field at sea level, particles of both signs occur in nearly equal numbers. The particles observed in these experiments are predominantly penetrating particles. It is therefore clear that these penetrating particles cannot be identified with the primary particles incident on the earth.

(2) The number of primary cosmic rays entering the atmosphere, calculated according to the method of § 4·2, is not sufficient to provide both the electronic rays in the upper atmosphere and the penetrating rays in the lower atmosphere.

This argument may be considered quantitatively thus*: at the equator the minimum energy for a charged particle to penetrate the earth's magnetic field is $15 \cdot 10^9$ e.v. (This is the minimum value allowed for vertical incidence at the earth's surface and is here taken as a mean for all angles.) The integration of Bowen, Millikan and Neher's curve shows that $3 \cdot 10^7$ ion pairs are produced per sq. cm. per sec. in the whole depth of the atmosphere, and this corresponds to a total energy of 10^9 e.v. per sq. cm. per sec. taking the energy per ion pair as 32 e.v. The number of particles entering the top of the atmosphere is then

$$\frac{10^9}{15 \cdot 10^9} = 0{\cdot}065 \text{ per sq. cm. per sec.}$$

as an upper limit, and, allowing that the incident particles have an energy distribution with a mean above $15 \cdot 10^9$, the number is probably about

$$0{\cdot}03 \text{ per sq. cm. per sec.} = 1{\cdot}9 \text{ per sq. cm. per min.}$$

But the number of penetrating particles at sea level is known to be nearly 1 per sq. cm. per min., and if we use as the absorption coefficient for hard radiation ($0{\cdot}08$ m.$^{-1}$ water), we find that the

* Data from T. H. Johnson, *Phys. Rev.* **53**, 499, 1938.

number of penetrating particles at the top of the atmosphere is about 2·2 per sq. cm. per min., and therefore greater than the total number of incident particles. In reality, the incident particles must give rise, not only to the *penetrating* particles, but also to the *electrons*, which multiply according to the curve of Fig. 6·5.

The argument given above indicates that the mesotrons must be produced in the atmosphere by some multiplication process. The incident particles may be electrons or protons, and the positive excess of incident particles suggests the latter.

7·7. Possible processes for the production of mesotrons.

If we consider the production of mesotrons from the theoretical standpoint, a number of processes appear to be possible (10). It is, however, known that the theory of the interaction between mesotrons and nuclear particles involves grave internal inconsistencies when the energies involved are greater than the rest energy of the mesotron. The theory may be used to suggest possible processes in a qualitative way but no significance can be attached to the estimates of probabilities.

(*a*) A light quantum of high energy can be absorbed by a nuclear proton, giving a neutron and a mesotron, or by a neutron giving a proton and a mesotron of the opposite sign:

$$h\nu + P \rightarrow N + U^{+},$$
$$h\nu + N \rightarrow P + U^{-}.$$

The probability of these processes has been calculated approximately for quantum energies about m_0c^2 ($\sim 10^8$ e.v.) and it appears that the number of light quanta absorbed in this way may be of the order of 2 per cent of the number absorbed by electron pair production.

(*b*) A light quantum of sufficient energy, in the neighbourhood of a nucleus, can give rise to a mesotron pair. The probability of this process is of the order $\left(\frac{m_0}{m}\right)^2$ times the probability of production of an electron pair, and is small compared with that considered in (*a*).

Heitler (11) calculates that the process (*a*) is capable of accounting for a large part of the mesotrons observed in the lower atmosphere. Direct experimental evidence of the production of mesotrons as secondaries could be obtained by measurements of the number of penetrating particles at high (balloon) altitudes. At the top of the atmosphere there should be only a small number of particles able to penetrate, say, 20 cm. lead, and the number of such particles should increase initially with depth in the atmosphere, reaching a maximum lower down than the maximum of the curve of § 4·1.

No process for the production of mesotrons from incident protons has yet been examined in detail from the theoretical side.

7·8. Multiple processes.

Heisenberg (10) found that the Fermi β-decay theory (with electron-neutrino exchange forces) contained the possibility of the creation of a number of electron pairs in a single act (explosion) when a very energetic electron entered an atomic nucleus. The Yukawa theory similarly suggests that an explosive shower containing mesotrons can be produced by the entry of an energetic mesotron or an energetic light quantum into a nucleus (11). Further, the particles in a heavy nucleus are bound together, so that some of the energy may be transferred to the nucleus, and result in the ejection of neutrons or protons (see chap. VIII). The theory does not give any definite probability for "explosion" processes, but suggests that they should be observable for incident energies of the order mc^2. It has been seen (§ 5·3) that there is some experimental evidence for these explosive showers, but that they appear to be rare.

Heisenberg and Euler (12) suggest that explosions are responsible for a large part of the bursts, particularly in the case of light elements. They show that the relative frequency of large bursts in light and in heavy elements as observed by Nie, does not accord with the cascade theory. The frequency of explosions actually observed in the expansion chamber does not, however, seem adequate for this purpose, and it is probable that the burst observations will receive another explanation. If the multiple processes, giving rise to showers of mesotrons, occur in sufficient number in the upper atmosphere, they may of course account for a considerable part of the secondary mesotrons. Experimental evidence is here lacking.

7·9. Electrons secondary to the hard rays.

Electrons occur in the cosmic rays at the bottom of the atmosphere in numbers much too large to be accounted for by the primary electronic rays, which are absorbed in accordance with quantum theory in the upper atmosphere. The latitude effect for the soft component, which is nearly the same as for the hard component, shows that soft rays at the bottom of the atmosphere can be produced by primary particles with energies as low as $3 \cdot 10^9$ e.v. Primary electrons of this energy can produce only insignificant effects through the thickness of the atmosphere (§ 6·8). It is therefore reasonable to consider many of the electrons observed at sea level as secondary to the hard component.

Evidence for this view is provided by the fact that showers have been detected at depths below sea level equivalent to 700 m. of water, and that at depths equivalent to 60 m. of water, cloud-chamber photographs show that the showers are mainly of the cascade type. Further, the frequency of showers in these experiments below sea level varies with the depth so as to remain nearly proportional to the intensity of the general radiation, while in the upper atmosphere the rise in shower intensity is more rapid than that of the general radiation (§ 5·6). It therefore appears that the shower-producing rays below the bottom of the atmosphere are in equilibrium with the general radiation, i.e. with the penetrating particles.

The following processes are suggested by theory as possible generators of secondary rays:

(1) The spontaneous decay of the mesotrons, already considered in § 7·4, gives rise to electrons, and these will multiply to some extent by the cascade process. Since the spontaneous

decay of the mesotrons is independent of the material traversed, while the absorption of the electrons increases with the density of the material, the number of decay electrons associated with a given number of mesotrons will be greater in the open atmosphere than under layers of denser matter.

Euler and Heisenberg have calculated the electron intensity at the bottom of the atmosphere, to be expected from mesotron decay. Taking the half-lifetime of the stationary mesotron as $2 . 10^{-6}$ sec. they find a small electronic intensity (2 per cent of the number of hard particles) due to the mesotrons which have been brought to rest before decaying, and a larger contribution due to mesotrons which have decayed while still moving rapidly. The electronic intensity at sea level (about $\frac{1}{3}$ electron per hard particle) can be accounted for by this calculation. Under water or other relatively dense material, the number of decay electrons becomes negligible.

(2) The mesotrons can give energy to electrons by direct collision, as is indicated by the ionisation they produce. Occasionally electrons will receive quite high energies by this process, and will then multiply by the cascade process. Bhabha has calculated the number of electrons knocked-on in this way, which accompany each mesotron:

TABLE 7·3

Electrons knocked-on by collision, accompanying each particle of mass $\sim 200m_0$.

(Bhabha, *Proc. Roy. Soc.* **164**, 257, 1938)

Primary energy e.v.	Electrons in lead		Electrons in air or water	
	Energy > E_c	Total	Energy > E_c	Total
10^8	—	—	—	—
10^{10}	0·09	0·19	0·03	0·07
10^{12}	0·16	0·34	0·07	0·15

The table gives separately the number of electrons with energies greater than the critical energy E_c, for which radiation loss becomes equal to the collision loss of energy (§ 6·6). $E_c \sim 10^7$ e.v. for lead, $1{\cdot}5 \,.\, 10^8$ e.v. for air or water, and particles with greater energy than this can give rise to cascades.

It is probable that both these processes are responsible for secondary electrons in the lower atmosphere, and that the Bhabha collision process is responsible for most of the electronic phenomena observed in underground and underwater measurements. Accurate data for the quantitative consideration of these processes are not yet available, either on the theoretical or experimental side.

Theory suggests also that the mesotrons may have a specific interaction with nuclear particles—the interaction being connected with the Yukawa field rather than with the electrical charge of the particles.

The theory shows severe internal difficulties if the energy of the particles is greater than the rest energy of the mesotrons. Its predictions must therefore be treated as suggestions of possible processes, and no importance may be attached to the estimates of their probability at the higher energies.

The mesotrons may interact with nuclear protons and neutrons, producing energetic quanta:

$$U^+ + N \rightarrow P + h\nu,$$

$$U^- + P \rightarrow N + h\nu.$$

These processes are the converse of those suggested for the birth of mesotrons in § 7·7. The light quantum produced will of course give rise to a cascade shower. The probability of the process has been calculated from theory for energies of the order 10^8 volts, and gives for the mesotrons a free path of about 2 m. in lead. If this result is applied to the atmosphere, together with cascade multiplication, it seems capable of accounting for the electronic component at sea level. There is as yet no experimental evidence for the reality of this process, and the comparative rarity of protons and neutrons (§ 7·1) in the lower atmosphere suggests that it does not happen as frequently as is here suggested.

REFERENCES

(1) Montgomery and Montgomery, *Phys. Rev.* **50**, 975, 1936.
(2) Anderson and Neddermeyer, *Phys. Rev.* **51**, 884, 1937.
(3) Corson and Brode, *Phys. Rev.* **53**, 77, 1938.
(4) Yukawa, *Proc. Phys. Math. Soc., Japan*, **17**, 48, 1935.
(5) Yukawa and Sakata, *Proc. Phys. Math. Soc., Japan*, **19**, 1084, 1937.
(6) Yukawa and others, *Proc. Phys. Math. Soc., Japan*, **20**, 319, 720, 1938.
(7) Bhabha, *Proc. Roy. Soc.* **166**, 501, 1938.
(8) Auger, Ehrenfest, Fréon and Fournier, *Compt. rend.* **204**, 257, 1937.
(9) Blackett, *Nature*, **142**, 992, 1938; Rossi, *ibid.*
(10) Heisenberg, *Z. Physik*, **101**, 533, 1936.
(11) Heitler, *Proc. Roy. Soc.* **166**, 529, 1938.
(12) Heisenberg and Euler, *Ergeb. exakt. Naturwissenschaften*, **17**, 1, 1938.

Chapter VIII

HEAVY SECONDARY PARTICLES IN THE COSMIC RAYS

8·1. Heavy secondary particles in cloud-chamber experiments.

Particles which can be distinguished by their specific ionisation as having masses comparable with protons are relatively rare in the cosmic rays. Apart from the possibility that some of the hard rays are protons (§7·1), there is some evidence for secondary heavy particles resulting from nuclear disintegrations brought about by the cosmic rays.

Heavy particles have been photographed in the cloud chamber, especially at high mountain altitudes (1, 2). Their tracks may be distinguished from those of contamination α-particles by their long range. In most cases they are associated with other tracks.

Observations on these tracks show that

(1) The majority of the tracks appear to be due to protons.

(2) These particles appear to go in all directions, and in particular the vertical direction is not specially favoured.

(3) The frequency of the tracks increases rapidly with altitude (Table 8·1 *b*).

(4) In a number of cases, several heavy tracks appear to diverge from a single centre, and these disintegrations have the following characters:

(*a*) The disintegrations appear to be produced in some cases by incident ionising rays and in some cases by non-ionising links.

(*b*) In some cases electron (thin) and proton (heavy) tracks appear to originate in the same centre.

8·2. Tracks observed in photographic emulsions.

Besides these cloud-chamber photographs, the tracks of strongly ionising particles have been revealed by microscopic examination of photographic emulsions. Certain photographic emulsions are remarkably free from the spontaneous appearance of developable grains. Plates coated with such emulsions, exposed for some months to the cosmic rays, show after development rows of blackened grains which correspond to the tracks of ionising particles in the emulsion itself. Similar tracks can be produced by α-particles or protons in the laboratory, but the cosmic-ray tracks are frequently longer than those of radioactive particles, and there is some evidence from the sparser distribution of black grains that they are produced by protons of high energy. Several workers have found that the number of tracks is greatly increased by the presence of a layer of paraffin, so that there is evidence for the existence of neutrons which produce fast protons by recoil. Further evidence for neutrons has been obtained by using emulsions containing boron, in which slow neutrons produce characteristic α-particle tracks by nuclear disintegration, and by the use of boron-lined electrical counters.

Table 8·1 *a* gives the increase with altitude in the number of proton tracks in plates covered with 1 mm. paraffin. In addition to the traces of single particles, Blau and Wambacher (3) have found in their emulsions stars of 3–12 tracks with a common origin.

TABLE 8·1

(a) Frequency of occurrence of heavy tracks in photographic emulsions covered with 1 mm. paraffin

(Schopper (4))

Height, km.	0·2	3·4	18
Tracks per sq. cm. per hr.	0·07	0·28	5·1
Ratio	1	4	73
Ratio for showers	1	5·5	—

(b) Frequency of heavy tracks in expansion chamber

(Anderson and Neddermeyer)

Height, km.	0	4·3
Ratio of number of tracks	1	11
Ratio of showers	1	8·5

Table 8·1 gives the variation of the frequency of heavy particles with altitude, as far as it is known. The variation is not known for the "stars" of associated tracks, but they probably increase quite rapidly with height.

It may be seen that the frequency of heavy particles increases more rapidly with altitude than the general cosmic radiation, and shows a rough correspondence with the increase of showers.

It therefore seems probable that these particles arise from nuclear disintegration by electrons or light quanta, and this view is supported by the observation of Anderson and Neddermeyer that the particles are often associated with electronic showers.

According to the ideas of Bohr, energy communicated to a nuclear particle may be divided up between the nuclear constituents, the effect being a general "heating" of the nucleus. Heisenberg has shown that a proton or neutron of moderate energy ($\sim 10^8$ e.v.), striking a nucleus, may transfer most of its energy to the nuclear constituents, and the nucleus thus heated may "evaporate", with the emission of several protons or neutrons. The appearance of some at least of the disintegration stars found in photographic emulsions is consistent with this view.

REFERENCES

(1) Anderson and Neddermeyer, *Phys. Rev.* **50**, 263, 1936.
(2) Brode, Macpherson and Starr, *Phys. Rev.* **50**, 581, 1936.
(3) Blau and Wambacher, *Wien Ber.* **146**, 623, 1937.
(4) Schopper, *Naturwiss.* **25**, 557, 1937.

INDEX